The DOG

First published in the UK in 2018 by

Ivy Press

An imprint of The Quarto Group
The Old Brewery, 6 Blundell Street
London N7 9BH, United Kingdom
T (0)20 7700 6700 **F** (0)20 7700 8066

www.QuartoKnows.com

British Library Cataloguing-in-Publication Data
A catalogue record for this book is available
from the British Library

ISBN: 978-1-78240-562-7

This book was conceived, designed and produced by
Ivy Press
58 West Street, Brighton BN1 2RA, United Kingdom

Publisher Susan Kelly
Creative Director Michael Whitehead
Editorial Director Tom Kitch
Art Director James Lawrence
Commissioning Editor Sophie Collins
Project Editor Joanna Bentley
Design JC Lanaway
Picture Researcher Katie Greenwood
Illustrator John Woodcock

Printed in China

10 9 8 7 6 5 4 3 2 1

MIX
Paper from
responsible sources
FSC® C008047

The DOG

A Natural History

ÁDÁM MIKLÓSI WITH
TAMÁS FARAGÓ, CLAUDIA FUGAZZA,
MÁRTA GÁCSI, ENIKŐ KUBINYI,
PÉTER PONGRÁCZ & JÓZSEF TOPÁL

IVY PRESS

Contents —◎

Introducing the Dog

For some people, dogs are colleagues who assist them in their jobs; for mavericks, they represent wolves, true reminders of the wilderness, and for city-dwellers, dogs may appear as demanding children, in need of constant care and devotion. There are many ways in which humans relate to these four-legged creatures . . . and, indeed, dogs come in so many shapes and forms that it is no wonder that people may sometimes get confused about the different roles they play.

In this book, we portray the dog both as an animal with a unique evolutionary history and as man's (and woman's) best friend. Our task has not been an easy one. There are so many expectations: Everyone, even those who are not dog owners, seems to be an expert on dogs. And the myriad of wonderful and heart-warming stories and anecdotes about them often hinder our objectivity.

There are many ways to characterize humans' relationship with dogs, especially the family pets with which many of us share our gardens, apartments, or even beds. People have the right to show their emotions when referring to their pet or companion dog as "my darling" or "my sweetheart," but science should also have its say. Dogs need to be respected as a species with its own destiny and allowed to be what it has become: a dog. Thus, respecting a dog as a friend is perhaps the best approach to the human–canine relationship. Friends can be attached to each other for all their lives but they are also able to lead independent lives for a shorter or longer time if circumstances require it. They help each other but do not expect an immediate return for favors. Friends enjoy being together just for the sake of it, but they also respect one another, allowing each other to develop independent personalities.

Right *Joint activities contribute to the health and wellbeing of both dogs and their owners.*

A RICHLY VARIED SPECIES

Canines are one of the most exciting groups of mammals on our planet. They come and go on the evolutionary stage, both in terms of time and space. At the start of the twenty-first century, at a range of locations across the Northern Hemisphere, wolves were on the brink of extinction. Now, they are back in many countries across Europe and also in the United States. However, life is never the same—evolution cannot repeat itself. These modern wolves also hybridize with coyotes and free-ranging dogs, possibly giving way to new forms of canine. In Europe, hunters had not seen the golden jackal for a century, but now, within the last ten years, jackals have reconquered old territories and ventured into new ones. Some of them have been reported hunting in the north of Europe, close to the Baltic Sea.

The existence of dogs and their many varieties is one of the most extraordinary proofs for evolution. Charles Darwin himself referred to domestic animals and especially to dogs when citing animal examples of evolution. However, as change is part of evolution, we should not expect the variation we have in our dogs today to stay with us forever. New times and new challenges may prompt the evolution of new creatures, and dogs are no exception.

While mutual friendship between dogs and humans may exist in billions of households around the world, in many situations we still want to be in charge. Humans can be quite a nuisance in this respect. One such case is dog breeding. Reproduction is a key to the evolution of a species, and any major failure can have fatal consequences in the long run. Especially in the case of purebred dogs, which are close to the hearts of many people, present breeding practices need to

Below *Despite having been domesticated, dogs often remind us of their wild relatives.*

be rethought. Neither irresponsible neutering nor arranged mating with a few males or the "perfect" champion male is advantageous for any breed. It may lead to a fatally reduced breeding population, the increase of inbreeding, and the emergence of physical malformations, illnesses, and behavioral problems.

As so many of us now live in cities, dogs may be one of our few connections to nature, so we should make every effort to keep them as healthy as possible and offer them the best life while allowing them to express their full biological potential. Dogs should be kept as companions only if the owner has the time and devotion to allow them the freedom of being a dog in addition to being a member of a family or other social community of humans. In this sense, dogs should be seen as the "wolves of the cities"—independent whether they are big or small, like to bark, or roam free in our green spaces.

Let's allow dogs to work if they enjoy it. People may or may not like to work, but dogs are different. They have been selected to like working with people, participating in joint activities. Research has also shown that many dogs are keen to work for people's "love," social feedback, and for the feeling that they are part of the family. As well as being genetic, as in the case of working breeds, this tendency can be facilitated through dog training. Thus, a well-trained dog, which has been chosen for this task, enjoys interacting with its owner. They would probably suffer if they were prevented from doing so. For dogs, working is closer to some kind of social engagement than a form of hard labor. In exchange, people express their feelings toward their dogs. But we should be careful not to demand too much; dogs also deserve to be dogs.

ABOUT THIS BOOK

Below *Dogs, especially those living in cities, need a lot of exercise to have a good quality of life.*

In this book, we hope to show you the dog from many different perspectives. Dogs are descendants of extinct wolflike canines, so they share many features with their wild cousins. Dogs also have a long and specific history with humans, and generations of dogs have witnessed how our societies have changed in the last 3,000–4,000 years. And, despite the fact that our relationship with dogs has become more intimate in some ways, dogs still remain dogs, in a good sense. So, we need to know about their biology: how they see, hear, and smell, and how they interact with one other and with humans, showing a wide array of sophisticated behavioral signals for communication. Dog owners have to become aware of the mental abilities of their companions in order to provide them with the necessary challenges to keep their minds sharp and active. This also ensures that dogs have a good quality of life during aging, as a more experienced and skillful dog has a lower chance of showing cognitive decline when it gets older.

Knowing about the development of dogs as puppies is crucial because this is the time when dog owners and breeders can have a huge influence on the future character of the dog. In sharp contrast to humans, who develop for around 18 years, in dogs maturation is much shorter, only one to two years. What a young dog may learn spontaneously after a few incidents may take much longer for an adult dog to acquire. Puppies learn as soon they are born, and if something is learned early, this can be remembered for their whole life.

And what about our future with dogs? In recent years, our societies have been changing at a rocketing speed. So far, dogs have been an exceptional means of providing us with a unique experience of friendship but now there are new competitors on the horizon. Television, the internet, and cell phones are giving many people, especially the young, the sense of being members of a community and there seems to be less time for developing human–canine relationships within the home. In the industrialized countries, the population of family dog numbers is stagnating or on the decrease—is this a sign of a relationship in decline?

Who can tell the future? But, for sure, humans have some responsibility for their creatures. The future of dogs lies in their behavioral flexibility, their ability to adapt to the newly emerging human needs in modern society. The new roles dogs play in our society give rise to new challenges for both dogs and dog trainers. We all have to make sure that dogs' needs are met, so they will continue to give us their company for centuries to come.

We hope that this book, which includes some of the newest insights from dog science, will help you, the reader, to respect your companion even more, or encourage you to find one of these wonderful partners to share your life with.

CHAPTER 1

Evolution & Ecology

Where Dogs Come From

Maned wolf

Bush dog

There is a rather striking resemblance between the appearance of any extant member of family Canidae (the group of carnivores closely related to and including dogs) and the long-ago extinct *Miacis*, the common ancestor of terrestrial predators. Thus, the Canidae show ancient anatomical features, or rather they are similar to the ancient form. This does not mean that the shape and functions of dogs and their closest relatives are obsolete—the high number of species still existing testifies just the opposite: The ancient form is still successful.

ORIGINS ON THE AMERICAN CONTINENT

The history of carnivorous mammals started about 55 million years ago (mya), not so long after the last of the great dinosaurs had disappeared at the end of the Cretaceous. Interestingly, *Miacis* emerged in North America—and the larger part of the evolution of Canidae also happened there. In the Paleocene (about 50 mya) the two main divisions of carnivores diverged, forming the catlike feliforms and the wolflike caniforms.

Toward the end of the Paleocene, about 34 mya, the *Caninae* subfamily appeared, and this would become the

Right Canids can look considerably different. The maned wolf is the tallest of all (3 feet / 0.9 m at the withers). The stocky bush dog is no bigger than a dachshund. Both live in South America.

Miacis

only surviving subfamily of the Canidae—and the ancestor of all the extant species of foxes, jackals, and wolves. The secret for their success may be that they were not overtly restricted to the hypercarnivorous ("meat only") diet of other subfamilies, which became extinct because of their narrow ecological tolerance of environmental changes.

Above The Miacis, *a primitive carnivore, populated both Eurasia and the North American continent about 55-33 million years ago. Animals like this were the ancestors of extant canids, bears, and weasels.*

"CANINE RADIATION" & THE COLONIZATION OF OTHER CONTINENTS

The evolution of early canids continued on the North American continent through the whole Oligocene until the second half of the Miocene. The so-called "Canine radiation" was an evolutionary "explosion" about 11 mya, when three major forms of canids—the wolflike *Canis*, the foxlike *Vulpes*, and the also foxlike *Urocyon* genera—appeared and became abundant in southwest North America (9–10 mya). Their success was hallmarked by the evolution of carnassials—a paired set of scissor-like molars and premolars in the upper and lower jaws, allowing the animal to perform an effective shearing bite—thus a better utilization of food.

These "modern" canids were those forms that left the North American cradle of terrestrial predators—first, about 8 mya through the temporarily available Beringian land bridge (between Alaska and the Kamchatka peninsula) toward Eurasia and Africa. Most extant species of wolves, jackals, and foxes evolved in the Old World after this exodus. The second major radiation of Canidae took place at about 3 mya, when the Isthmus of Panama formed. This allowed some of the North American species to invade South America, where, besides the gray fox (*Urocyon cinereoargenteus*), endemic species evolved such as the bush dog (*Speothos venaticus*) and the maned wolf (*Chrysocyon brachyurus*).

THE PLEISTOCENE & MODERN-DAY DISTRIBUTION OF CANIDS

Commonly known as the Ice Age, the Pleistocene is characterized by repeated cold (glacial) and warm (interglacial) periods, starting about 2.8 mya and ending about 12,000 years ago. For Canidae, the notable events during the Pleistocene were the repeated waves of colonizing species (especially jackals) invading Africa from the north, and the arrival of Eurasian gray wolves back to their "ancestral land," North America.

Below *The first canids emerged in North America about 40 mya. Their descendants arrived in Eurasia only around 8 mya. Closer relatives of dogs (genus* Canis*) evolved in the Old World. Wolves "returned" to America less than 1 mya.*

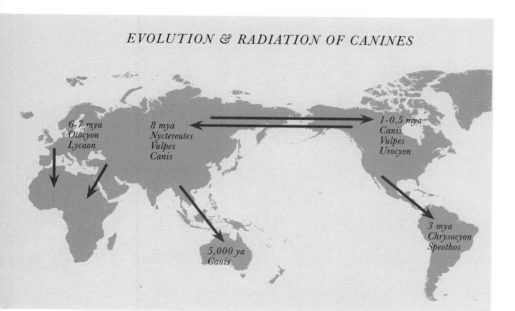

EVOLUTION & RADIATION OF CANINES

*6–7 mya
Otocyon
Lycaon*

*8 mya
Nyctereutes
Vulpes
Canis*

*1–0.5 mya
Canis
Vulpes
Urocyon*

*5,000 ya
Canis*

*3 mya
Chrysocyon
Speothos*

Why Caninae Survived ～ℓ

With evolution comes not only the emergence of new species but also the extinction of many others. Extant canid species around in the world provide proof that this group of predators can be considered highly successful, especially because we know that many of their former distant or closer relatives have already disappeared from the Earth. Besides the larger impact of geological and climatic changes, the survival and extinction in the case of Caninae was mostly a question of their ecology.

BASIC ECOLOGY OF THE EXTANT CANIDS

Depending on their body size, canids can consume prey as small as insects, or subdue large animals such as the elk and moose. However, almost all extant canid species are not typical hypercarnivorous species because they obtain only about 70 percent of their diet from animal protein sources; the rest comes from eating plants, fruits, or even nuts.

Canids are probably never solitary, maintaining at least a loose pair bond almost all year around—but more commonly they live in pairs, smaller families, and, in the case of some species, in larger packs. Extant canids can all be considered as social species; the gray wolf, the dhole, and the African wild dog may also fit the term "hypersocial."

Finally, canids can be regarded as very devoted parents, spending relatively long periods yearly on raising their offspring. Depending on their social habits, both adults and the previous year's offspring participate in tending the young, which are born small, blind, and helpless and require a long period of parental care.

THE SURVIVAL OF CANIDS AT THE END OF THE PLEISTOCENE

The end of the last glacial in the late Pleistocene came with a whole wave of extinctions, called the disappearance of the Ice Age megafauna. Many hundreds of large terrestrial mammalian species went extinct in a relatively short period of time, including successful survivors of several recurrent glacials and interglacials such as the woolly mammoth and the saber-toothed tiger. Canids were among those animals that fared much better than other taxa, with perhaps only one famous exception—the dire wolf (*Canis dirus*), which was one of the hallmark species that disappeared at the end of the Pleistocene.

THE CASE OF THE DIRE WOLF

The dire wolf shows clearly the role of feeding ecology in the temporary success and later demise of a predator. Evolved on the American continent, dire wolves were the size of the largest of today's gray wolves—not especially big compared to their contemporary competitors such as the saber-toothed tiger, yet fully capable of hunting the largest prey species available in their time: bison, wild horse, and even mammoths. According to skeletal fossil records, dire wolves were highly social hunters that once populated both Americas. However, they started to show a population-wide decline about 20,000 years ago and went totally extinct during the next 10,000 years.

The main factor in their decline was their strong dependency on megaherbivore prey. Dire wolves were hypercarnivorous and could not flexibly switch to smaller prey when the largest herbivores became rarer. The gray wolf and the other extant canids survived because (besides their effective social behavior) they could react to the changing diet opportunities with greater flexibility.

Above *Dire wolves evolved in North America and became extinct only toward the end of the last Ice Age. Being specialized to hunt the largest available prey, they eventually lost out to the more flexible gray wolves.*

Left *The mighty woolly mammoth was a widespread inhabitant of the Northern hemisphere during the Ice Age. Its relatively recent extinction (about 10,000 years ago) was most probably caused by swift climatic changes.*

Bottom left *The gray wolf is one of the most successful Canids. This large carnivore lives, reproduces, and hunts in packs. A pack can subdue even large-hoofed prey such as moose and elk.*

Distant Relatives of Dogs

The *Canis* family is represented by a few very closely related species that inhabit all continents except Antarctica (see table below). The close relationship is supported not only by their similar form and comparable life history, but also by their genetic makeup, which is so similar that individuals from different species can breed with one other. Interbreeding may also happen in nature, providing the possibility of further evolution in this group of predators. This also means that the English names such as "wolf," "jackal," and "coyote" are based more on tradition rather than reflecting a biological category.

WOLVES (*CANIS LUPUS*)

After the major extinction wave at the end of the Pleistocene the wolf survived as the top predator in North American and Eurasia. While the gray wolf remained the most abundant species, it has evolved into many divergent subspecies that differ in size, food choice, and lifestyle. Most recently, the Ethiopian wolf (formerly Ethiopian jackal) was assigned to this group because researchers discovered it is genetically more closely related to wolves than jackals, despite living in Africa.

Above *The black-backed jackal (1 and 4), side-striped jackal (2), golden jackal (3), and Ethiopian wolf (5, formerly Ethiopian jackal, now regarded as a wolf species). Jackals live in various locations in Europe, Asia, and Africa.*

COMPARATIVE SUMMARY OF *CANIS* SPECIES (BASED ON SHELDON 1988)

Species	Shoulder height	Weight	Diet
Side-striped jackal (*Canis adustus*)	16–20 in./41–50 cm	14–31 lb./6.5–14 kg	Omnivorous; carrion, small animals, plants/fruits
Golden jackal (*Canis aureus*)	15–20 in./38–50 cm	15–33 lb./7–15 kg	Carrion, small animals; coop. hunting
Black-backed jackal (*Canis mesomelas*)	15–19 in./38–48 cm	13–30 lb./6–13.5 kg	Carrion, plants/fruits; coop. hunting
Ethiopian wolf (*Canis simensis*)	21–24 in./53–62 cm	24–42 lb./11–19 kg	Rodents; hunts alone
Gray wolf (*Canis lupus*)	18–32 in./45–80 cm	40–132 lb./18–60 kg	Carnivorous; carrion, plants/fruits; coop. hunting
Coyote (*Canis latrans*)	18–21 in./45–53 cm	15–44 lb./7–20 kg	Carnivorous; carrion, plants/fruits; coop. hunting
Red wolf (*Canis rufus*)	26–31 in./66–79 cm	35–90 lb./16–41 kg	Small animals, carrion, plants

COYOTES (*CANIS LATRANS*)

This species evolved in North America (and is an endemic species there). Coyotes have a very similar lifestyle to wolves, although they are somewhat smaller and do not establish large groups. Until recent times coyotes have occupied only the more southerly areas of North America; but recently the populations have started to migrate northward.

JACKALS

This group within *Canis* is subdivided into three species. Jackals are generally much smaller than wolves and coyotes, and tend to live in small family groups. The golden jackal (*Canis aureus*) is commonly distributed in Southern Europe and Southern Asia, but recently this species started to migrate to the north of the European continent—for example, jackals were sighted in Estonia in 2013. The black-backed jackal (*Canis mesomelas*) and the side-striped jackal (*Canis adustus*) inhabit Africa from the Sahara south. They prefer to live in open areas and move around in pairs or small families.

RED WOLF & OTHER FORMS

Researchers disagree about the phylogenetic status of some wild canine populations. For example, the majority consider the red wolf as a separate species (*Canis rufus*), but it could represent a hybrid between wolves and coyotes. More recent findings also indicate that wolves and coyotes, and even free-roaming dogs, may hybridize more frequently than thought previously. This may lead to specific populations of canines being more successful in surviving because they are better able to tolerate ecological changes, including increasing temperatures and the continuous and broadening threat of human disturbance.

Below *Species of the Canis genus are closely related and this is indicated by similarities in form and behavior. The gray wolf (6 and 7) has less than a handful of relatives: the red wolf (8) and coyote (9).*

Gestation, litter size & care	Social organization	Home range
57–70 days (max. 7 offspring)	Pair + offspring	ca. 0.4 mi.²/1.1 km²
63 days (max. 9 offspring); biparental, alloparental	Very variable, pair + offspring (+ yearlings)	Hunting range 1–7.7 mi.²/2.5–20 km²
61 days (max. 9 offspring); biparental, alloparental	Pair + offspring	ca. 7 mi.²/18 km²
60–2 days (max. 6 offspring); biparental, alloparental	Pair + offspring	ca. 1.5–2.3 mi.²/4–6 km²
62–5 days (max. 13 offspring); biparental, alloparental	Very variable, pair + offspring + yearlings	ca. 7–5,000 mi.²/18–13,000 km²
ca. 60 days (max. 12 offspring); biparental, alloparental	Very variable, pair + offspring + yearlings	ca. 7–39 mi.²/1–100 km²
60–2 days (max. 8 offspring); biparental	Very variable, pair + offspring + yearlings	ca. 15–31 mi.²/40–80 km²

Emergence of Wolves

THE IMPORTANCE OF GRAY WOLVES

Gray wolves (*Canis lupus*) are probably the best-known, most iconic wild canid species. Besides being the closest living relative to the dog, the gray wolf is the most successful large terrestrial predator of recent times. As apex carnivores, with the exception of humans, wolves probably had the greatest biological impact on the late evolution of the ecosystem of the Northern Hemisphere.

MAIN STEPS OF EVOLUTION IN A NUTSHELL

The gray wolf is a relatively young species, although wolflike members of the genus *Canis* were abundant in the late Pliocene and most of the Pleistocene epoch (3–1 mya). It evolved in Eurasia, in several genetically distinct clades. Besides inhabiting the whole of the enormous Old World (except Africa), gray wolves appeared for the first time in North America relatively recently, less than 300,000 years ago.

Below *Although gray wolves can be found all around the Northern hemisphere, compared to their vast home range a few thousand years ago, their recent distribution shows a strong decline due to their conflicts with humans.*

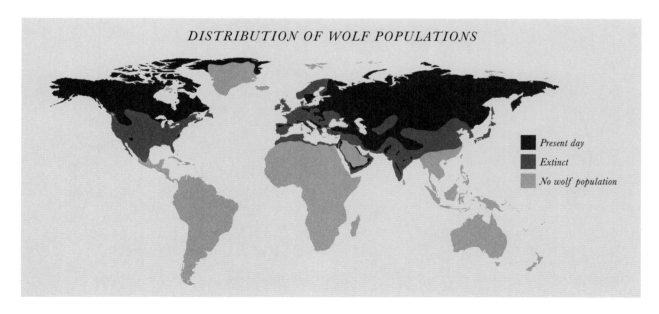

DISTRIBUTION OF WOLF POPULATIONS

Present day

Extinct

No wolf population

Paleontology and molecular genetics confirmed that from today's existing members of the family *Canis* the earliest diverged line led to the golden jackal. Somewhat surprisingly, the coyote is almost uniformly considered as the closest living relative to the gray wolf because other, more closely related species died out in Eurasia.

WOLVES IN THE ICE AGE

At the peak of the Ice Age (15–25 thousand years ago), permanent ice covered most of the North American continent to the south of the Great Lakes, large areas of Eurasia (mostly modern-day Russia), the whole of Scandinavia and the British Isles, and Europe down to the northern Carpathians. The "cradle" of the species was assumed to be in the eastern part of Eurasia. Beringia (an area of land joining Alaska and the easternmost end of Eurasia) remained ice-free and the Bering Strait offered a possible route for early wolves and their relatives to move back and forth between the "old" and "new" worlds. The last and most successful wolf-invasion occurred only 80,000 years ago. As the last ice age was probably the fiercest of them all, further incursions of Eurasian wolves became impossible due to the melting of ice fields across the whole continent.

THE APEX PREDATOR OF THE NORTH

Although the gray wolf is the largest of all extant Canidae (large males of the holarctic type can reach 176 lb./80 kg in weight and 32–34 in./80–85 cm in height), its dimensions are still smallish compared to some of the bears, and especially to the great cats (such as the *Smilodon*) that went extinct just toward the end of the last glacial period.

The secret of the unique evolutionary success of wolves is their ability to eat various prey types and to form highly effective social groups. In times of need, wolves are able to survive on small prey (such as rabbits and rodents), although they mostly hunt the largest available prey—hoofed animals. Importantly, wolves are not large enough to be able to kill an adult moose or elk alone, but here is where the formation of large packs becomes an advantage. Wolves cooperate to kill a larger animal and share the prey among the pack members. They also raise their offspring as a communal effort.

Wolves were, and still are, very mobile and able to cover large areas, both as individual animals and as populations. This feature gave the species further advantage when it was possible to exploit new lands, or when it was time to retreat from worsening climatic conditions. The highly successful gray wolf became abundant in the postglacial Northern Hemisphere, where game was plentiful in the vast forests. Wolf populations reached their peak only a few thousand years ago, when they met their fate—humans.

Below *Wolves are highly social. They use various vocalizations, of which howling is undeniably the most well-known. Howls are used to synchronize pack activities and also to herald the presence of wolves to neighboring packs.*

HUMAN & WOLF—
WOLVES OF THE PRESENT

Modern humans (*Homo sapiens*) are even younger as a species than wolves. These human populations settled in Europe and Asia 40,000–60,000 years ago, when they first met various representatives of the extended wolf population. There is little evidence of hostility between man and wolf up until the last 10,000 years. Both species were highly skilled group hunters of larger and smaller prey, and did not regard each other as suitable prey. Although at some Paleolithic excavation sites the bones of wolves killed by humans were discovered, their amount did not exceed the level of incidental hunting. Hunter-gatherer human groups might eventually have become wolves' rivals, but there was not yet open warfare.

The situation changed when humans turned toward agriculture and livestock keeping. Wolves became the hated and feared predators of those animals that provided a living for our ancestors—therefore they had to be eradicated. Eventually, the "big bad wolf" became the embodiment of "evil"—and medieval people successfully exterminated most of the European wolf population. Another anthropogenic factor in wolf recession was the alteration of the landscape: Due to large-scale deforestation, wolves lost both their habitat and prey species. In Europe alone, at the beginning of the twentieth century, the proportion of forested land had declined to 20 percent from an estimated 75 percent at the turn of the tenth century.

Today the world's total wolf population is estimated at around 300,000 specimens. Gray wolves were almost entirely extirpated from their former habitats of southern North America, northern and western Europe, India, and Japan. Apart from isolated populations across these areas, the majority of gray wolves today live in the coldest and most forested regions of North America and Eurasia.

During the twentieth century wolves became an endangered species in many countries, and efforts were made for the reintroduction of wolves to particular sites. The best-known reintroduction took place in the Yellowstone National Park in the United States in the 1990s, with large-scale ecological consequences that mainly resulted in a decrease in the elk and deer populations and the recovery of forested areas due to the lighter grazing pressure. The wolf is now also expanding in Europe, with new populations established in Switzerland and Germany.

The preservation (and reintroduction) of wolves remains a sensitive issue everywhere and needs to be carefully considered before implementation. Wolves do not discriminate in their prey between game and livestock, so the coexistence of humans and wolves as two "top" predators is far from being settled. Intensive debates have arisen in countries, including in the United States, where, after many years of having been protected, the expanding wolf populations have made some policymakers argue for the reintroduction of wolf hunting.

Below *Today wolves can live mostly undisturbed only in the High North. This adaptable carnivore still lives in large numbers where winters are long, large prey is abundant, and humans are rare.*

First Links with Humans ✦

Domestication is an evolutionary process during which some ancient wolf populations became adapted to humans and to the anthropogenic environment through a series of genetic changes. The exact details of this process, however, have remained obscure, and this has kept many scientists busy in recent decades working on new theories and ideas. All agree on one point: The history of dogs and humans in the last 16,000 to 32,000 years has been tightly interwoven.

Left *A limestone statue from Cyprus from around the 4th to 3rd century* BCE.

Above *This Egyptian mummy contains dog bones and is believed to date from between 400* BCE *and 100* CE.

Below *Dogs feature on Ancient Greek black-figure work, such as this terracotta skyphos (drinking cup) from around 500* BCE.

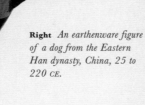

Right *An earthenware figure of a dog from the Eastern Han dynasty, China, 25 to 220* CE.

THEORIES OF DOMESTICATION

There are many theories of domestication, most of which have some credible elements. Considering all of them together probably gives the most plausible account of the sequence of events. Here are some examples:

1. Socializing wolf cubs (individual-based selection): Pups of wild canids show very diverse behavior toward humans, so it is possible that wolf cubs raised by humans and showing the "right" temperament were selected for over many generations.

2. Wolves domesticated themselves (population-based selection): Humans' activities (such as hunting) produced an excess of garbage, a novel, easy-to-exploit food source. This could have been utilized by (some) canine populations over generations. Smaller individuals, which could live on such food and were not frightened by the presence of humans, become gradually isolated from the rest of the wild population (just like city pigeons).

Below *According to archaeological records, ancient Arctic dwellers may have been among the first humans to breed dogs for pulling sleds.*

3. Preference for wolves (human group selection): Human groups with an affiliative tendency toward canines had selective advantage, because observing their behavior might have helped in hunting and in establishing settlements. As a result, both humans attracted to dogs and dogs themselves become widespread.

4. Diversification of dog roles: Originally dogs had only restricted roles, but later humans found ways to employ them in different tasks (such as hunting partner, heater, guard, sled-puller, and food source).

HUMAN MEETS WOLF

300,000–400,000 years ago

Three species of the *Homo* genus, who had left Africa earlier, probably encountered wolves along their journey. However, no change in wolf populations took place during this time.

45,000–120,000 years ago

Modern humans, *Homo sapiens*, left Africa and colonized Europe and East Asia in several waves. Recent dogs may have emerged as a consequence of encounters between modern humans and wolflike canines. Remains of an Upper Paleolithic doglike animal were discovered in Belgium, and found to be about 31,000 years old.

10,000–12,000 years ago

Remains suggest that dogs' size reduced by 38–46 percent, and they participated in hunting. In an Israeli burial site the hand of a deceased human was positioned over the body of a puppy, suggesting an affectionate relationship. People practiced ritual burials of dogs in all parts of the world; other domesticated species were buried much less frequently.

8,000–10,000 years ago

The presence of dogs is confirmed by wall drawings depicting hunting scenes in Turkey. Small-bodied dogs have been recovered in Germany, Sweden, Denmark, Estonia, and England. In a Serbian site on the bank of the Danube, a high number of broken dog bones and skulls suggest that fishing and hunting communities ate these animals. The first archaeological evidence that dogs reached North America dates to 9,000 before the present.

4,000–6,000 years ago

Many drawings and sculptures depicted dogs. In parallel with rapid technical changes, humans started to select dogs for various working roles, which resulted in characteristic sets of morphological and probably behavioral traits. On Egyptian pottery and rock art most dogs look like sight hounds with slender bodies, erect ears, and curly tails.

3,000–4,000 years ago

Animal figures and rock carvings suggest that dogs were used in herding and also in guarding. Individuals varied in size and had a curled tail and floppy ears. The first dogs arrived in Australia and the free-ranging populations evolved to dingoes. Dingoes still have a prominent role in the culture of Aboriginal Australians, and they are depicted on rock carvings and cave paintings.

20,000 years ago

(after the last glacial) Human populations expanded and by 10,000–15,000 years ago most continents had human occupants. During this phase agriculture emerged in several places.

12,000–15,000 years ago

Humans established large permanent settlements, which provided a barrier between wild and anthropogenic canine populations. Clear evidence for human/canine cohabitation comes from Germany in the form of 13,000-year-old bones. Trading humans and dispersal events could have rapidly widened the distribution of doglike animals.

6,000–8,000 years ago

Dogs were introduced from the Near East to Egypt and later dispersed throughout Northern Africa. Joint burials of dogs and humans suggest an intimate relationship between Native American hunters and dogs. The most widespread dog was the Mesoamerican common dog (withers height 16 in./40 cm).

1,500–3,000 years ago

During the Roman period, selection for increased size is evident, but very small lapdogs also became more common. This suggests the appearance of targeted selective breeding for looks rather than for their value at work. During this time dogs reached the most southerly parts of Africa with the migrating Bantu peoples.

150–200 years ago

Most dog breeds were developed.

PRESENT-DAY TRADITIONAL SOCIETIES & THEIR DOGS

The Turkana people of Kenya have the highest prevalence of tapeworm infection in the world, probably due to the unique role dogs play in their life as nomadic pastoralists. Dogs are not only the playmates of children, but they also clean up after the child if it defecates or vomits. Dogs also lick cooking-ware and serving-ware clean, and consume the menses of the women. Given that the fresh water supply is limited in this semiarid region of northwest Kenya, this practice is understandable.

Who is the Ancestor of the Dog? ~&

During the 1990s molecular studies were beginning to be applied in archaeozoology, and in 1999 researchers suggested that dog domestication began at least 135,000 years ago. They reached this conclusion after comparing the number of genetic changes in the mitochondrial (mt) DNA obtained from wolves, dogs, and coyotes. According to these studies, dogs and wolves diverged more than 100,000 years ago. However, these assumptions have been widely contested, with newer studies providing more precise estimates about the date and place of domestication.

DO DOGS COME FROM ASIA, EUROPE . . . OR BOTH?

These later studies, using larger collections of mtDNA from different wolves and a wide range of dog breeds, as well as more genetic markers and more sophisticated statistical models, concluded that all dogs share a common ancestry with wolves from the region south of the Yangtze River in China. The population living in this area was more diverse genetically, which suggested that founders of other populations dispersed from here, and the descendants that migrated to other areas in Eurasia took with them only a subset of the original genetic diversity.

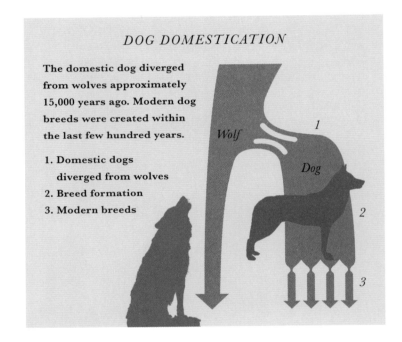

DOG DOMESTICATION

The domestic dog diverged from wolves approximately 15,000 years ago. Modern dog breeds were created within the last few hundred years.

1. Domestic dogs diverged from wolves
2. Breed formation
3. Modern breeds

Wolf

Dog

1

2

3

When researchers also included the DNA of free-roaming village dogs (which were considered to represent "ancestral" dog genomes because they had not been subjected to selective breeding and existed in isolation from other canids), then researchers concluded that dogs originated in Eurasia, possibly including North Africa (Egypt). Today, South Asia, Europe, and also the Near East are considered as potential sites of dog domestication.

WHEN DID DOMESTICATION TAKE PLACE?

An independent study reported that a 33,000-year-old fossil dog from Siberia (Russia) might have been related to modern dogs and prehistoric North American wolves but not to modern wolves. When more fossils were involved in the analysis, researchers found that modern dogs were related to ancient European canids, and the onset of domestication might have occurred 19,000–32,000 years ago.

Whole genome sequencing of individuals originating from China, the Near East, and Europe supported the view that domestication started 11,000–16,000 years ago, and numerous bottlenecks (sharp reductions in the population size) have occurred since then. The data suggested that the direct ancestor of dogs is now extinct. It is therefore impossible to localize the region of domestication based only on extant specimens.

TOWARD CONSENSUS

In 2016, nearly all researchers involved in studying dog domestication offered their canid samples for a mega-investigation. The results suggested that dogs were domesticated twice: approximately 15,000 years ago in Europe and 14,000 years ago in Asia. Between 6,400 and 14,000 years ago the population in Western Europe was largely replaced by Asian dogs. Thus, it is likely that domestication of dogs was initiated several times, but many proto-dog lineages failed to survive.

Above *Early dog populations could have regularly hybridized with wolves. Some dog genes also found their way into the wolf populations. Descendants, such as this wolf/dog hybrid, retain fragments of parental characteristics in unique combinations.*

THE FIRST COMPLETE GENOME OF AN ANCIENT DOG

The inner ear bone of a nearly 5,000-year-old dog unearthed in Ireland was used to obtain DNA for the first whole genome sequencing of a dog fossil. The ancient Irish dog's DNA was compared to the DNA of 605 modern dogs from around the world. The family tree of these animals revealed a divide between European dogs (such as the ancient dog from Ireland and the golden retriever) and Asian dogs (such as the shar-pei and village dogs from Tibet).

WHAT IS A SPECIES?

It is difficult to define the term "species" in a way that applies to all organisms. The biological definition of species states that individuals that can produce fertile offspring represent the same species.

According to the ecological species definition, a species consists of potentially interbreeding natural populations that are reproductively isolated from other such groups.

These definitions lead to some confusion because, in the case of the former, the dog should be categorized as a subspecies of wolf and named *Canis lupus familiaris*.

But in the case of the ecological species definition, dogs and wolves are two separate species because they do not typically hybridize in nature. Thus, the dog should retain its Latin name, *Canis familiaris*, which it was given by the famous Swedish scientist Carl Linnaeus. Nowadays it is quite unfortunate that both names are used in the scientific literature.

HYBRIDIZATION BETWEEN DOGS & OTHER MEMBERS OF THE *CANIS* GENUS

After more than a century of dispute it is now agreed that the closest living relative of the dog is the gray wolf (*Canis lupus*). There is no evidence of any other canine species contributing to dogs' genetic lineage. This is somewhat surprising because dogs can crossbreed with golden jackals and coyotes. However, up to now such hybrids have rarely emerged spontaneously in nature. But crossing different species of canines has been a practice for several thousand years, and hybridization has become more frequent in recent years:

• Coyotes, wolves, and dogs were deliberately crossbred in Pre-Columbian Mexico in order to obtain loyal, but resistant guard dogs.

Right (both) *Jackal-dog hybrids are hard to train, but after breeding back to dogs, Sulimov dogs have been bred with a superior sense of smell and are used for airline security in Russia.*

Opposite *Modern dog breeds, such as German shepherd dogs and golden retrievers, are the results of selective breeding and inbreeding for around 100 to 120 generations.*

- According to some accounts, northern Canada's Aboriginal peoples were mating both coyotes and wolves with their dogs to produce more resilient sled dogs.
- Crossbreeding in nature between wild jackals and dogs was first confirmed in Croatia in 2015. The Sulimov dog (or Shalaika) is a Russian jackal–dog hybrid developed for security as a sniffer dog.
- So-called wolfdogs (wolf–wolflike-dog hybrids) are popular companion dogs in the United States, despite the fact that these animals present a potential threat because they are less predictable and trainable and could become dangerous. Several recognized dog breeds exist that are the results of crossbreeding between wolves and dogs, including the Saarloos wolfdog and the Czechoslovakian wolfdog. The first generation was backcrossed to dogs (German shepherd dogs) to decrease the amount of wolf DNA.

DOMESTICATION FACTS

- Dogs are the oldest domesticated animal.

- The ancestors of modern dogs and wolves started to diverge morphologically 16,000–35,000 years ago. Considering the archeological evidence, 15,000 years appears to be a convenient date for dog domestication.

- Dogs do not come from Africa, America, Australia, or the Indian subcontinent.

- Domestication happened independently in Eastern and Western Eurasia, but Western Eurasian dogs were mostly replaced by the Eastern dogs.

- The wolf population that was the direct ancestor of the dog is now extinct.

Emergence of Modern Dog Breeds —℘

For a long time, those ancient dogs that were allowed to join human social groups did not develop into distinct types depending on function, but were only roughly selected based on some crucial behavioral traits related to their ability to adjust their actions to those of humans and avoid conflict situations. During the thousands of years in which dogs and humans have been living together, the selection criteria have followed constantly changing goals.

HOW ANCIENT ARE THE "ANCIENT" BREEDS?

Although many modern breeds look like those depicted in ancient artefacts, DNA data from today's representatives of these so-called "ancient" breeds provide evidence that most of them have little in common genetically with their ancestors. It has long been believed that these breeds had origins dating back over a thousand years. However, only a few have a somewhat closer relationship to wolves' genetic makeup. These breeds, for example, the shar-pei, akita inu, shiba inu, chow chow, and basenji are proved to be genetically separated from the overwhelming majority of modern breeds. In some cases (such as the pharaoh

hound), the original type was extinct and later a modern breed was created from many existing variants, and the new design retained only the outward appearance of the original form, not the genetic ancestry.

FOREVER YOUNG?

Some of the anatomical modifications of modern dog breeds have been dominated by paedomorphosis, so that the adult dog resembles a juvenile stage of the wolf. In a sample of dog breeds representing different degrees of physical similarity to the wolf (from the Cavalier King Charles spaniel to the Siberian husky), the signaling ability of ancestral dominant and submissive behavior patterns correlated positively with the degree to which the breed physically resembled the wolf. This suggests that during domestication of the modern breeds physical paedomorphism has been accompanied by changes in the signaling abilities.

Below *The Ancient Greeks and Romans used large dogs, such as the Molossian, as guards or military dogs.*

FUNCTION OF DOGS

In the broad sense, dog breeding means some sort of intentional selection of the parent generations and/or the offspring in order to achieve some desirable traits. Breeding dogs for particular characteristics (phenotypes), by selective mating and testing whether the offspring fit their function, had been going on for centuries before the fundamental principles of heredity were revealed. Artificial (human-directed) selection in dog breeding has influenced behavior, shape, and size, so that nowadays the dog probably has the widest variety of phenotypes among all species.

FIRST DOG TYPES

Only archaeozoological findings and information from drawings and paintings provide some hints about the emergence of the first breeds. It is very likely that there is little relationship between the dogs seen on the walls of the pyramids, or depicted in decorations of ceramics, and today's breeds. Thus, the heavily built ("mastiff-like") dogs used in wars by the Greeks and Romans may share their form with some breeds existing today but are probably not related genetically. The hound type is also one of the early dog forms that had existed as a reproductively isolated breed, but again the genetics of those may be different from present-day hounds originating from the Near East.

Above right and inset *In spite of its name, the pharaoh hound is a modern breed, without links to Ancient Egypt.*

Right *The akita is a Japanese breed that is genetically close to wolves.*

Free-ranging Dogs:
Stray, Village, Pariah Dogs

The level of dependency on humans of individual dogs appears to vary a lot today. In the industrialized countries of the world many dogs live as companions to humans. Meanwhile, large numbers of dogs exist around the world that seemingly do not have much in common with their pampered pet cousins.

STRAY, VILLAGE, PARIAH DOGS—DEGREES OF DEPENDENCY

While in the more industrialized parts of the world there is a clear line between "owned" and stray dogs (with the latter regarded as some kind of anomaly), villagers in Africa, for example, enjoy a rather mixed situation. Dogs can be categorized by two traits that describe dog–human relationship: *dependency* to survive and *restrictions* in reproduction.

Fully dependent, fully restricted— family dogs and working dogs

These dogs have dedicated owners, their activity and reproduction is supervised strictly by the owner, and their needs for nutrition, accommodation, and veterinary care are solely and directly provided by their human companions.

Fully dependent, partially restricted—stray dogs

In some industrialized societies, these dogs are in reality owned by someone who does not fully control the dog's roaming activity; or they can be "true strays" that were lost, abandoned, or relinquished by their owners. Being an ownerless stray is not a stable state for these dogs, and sooner or later these animals are usually reintegrated into the society by adoption or through dog shelters.

Above and below left *Free-ranging dogs are lean and muscular, and often show "wolf-like" features, such as pricked ears. In general, their anatomical attributes are advantageous for survival in a world where life is tough and human assistance is unlikely.*

Right *Free-ranging or pariah dogs are abundant in the cities and villages of India, Africa, and South America. These dogs show a striking resemblance to each other around the world.*

Partially dependent, not restricted— village or pariah dogs

Pariah dogs do not have a dedicated owner, although they utilize mostly human resources (especially for food and shelter). They are highly skilled survivors among these conditions, and, considering their reproductive potential (see below), can be regarded as one of the most typical ecological variant of dogs.

Not dependent, not restricted— feral dogs

This is a debatable category because there are only a few specific cases for which there is evidence that these dogs live in self-sustaining populations and are fully independent of human (food) resources. Some dogs living in Southeast Asia and Australasia (for example, singing dogs and dingo) are the best fit for this category.

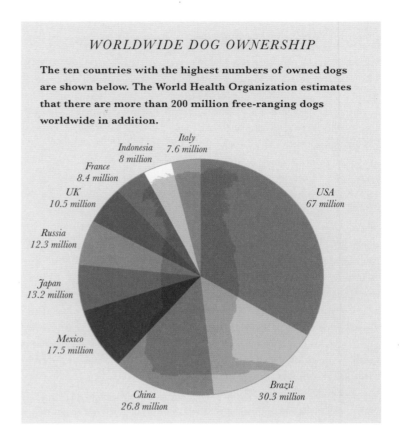

WORLDWIDE DOG OWNERSHIP

The ten countries with the highest numbers of owned dogs are shown below. The World Health Organization estimates that there are more than 200 million free-ranging dogs worldwide in addition.

Italy
7.6 million

Indonesia
8 million

France
8.4 million

UK
10.5 million

USA
67 million

Russia
12.3 million

Japan
13.2 million

Mexico
17.5 million

China
26.8 million

Brazil
30.3 million

FACING THE FACTS—
PARIAH DOGS ARE
UNIQUELY SUCCESSFUL

There are only estimates regarding the number of dogs living around the world. The boldest assumes that we share our planet with about 1 billion dogs, and most likely only about 20 percent of these live under close human supervision—which means that there are about 800 million pariah dogs worldwide. The vast majority of them live in warmer climates, especially in India and Southeast Asia, Africa, Mexico, and South America.

Pariah dogs lack the bewildering variability of looks we are familiar with in purebred dogs and their hybrids—indeed, pariah dogs show surprising uniformity across continents. They are small to medium-sized, short-haired dogs with rectangular-proportionate build, and mostly tan or tan-and-white. This suggests that pariah dogs have undergone natural selection that resulted in an economic but tough organism, highly successful in its ecological niche—that is, at the fringe of human society.

THE ECOLOGY &
BEHAVIOR OF PARIAH DOGS

Pariah dogs depend on human food resources—however, they are seldom provisioned willingly by humans, but rather fend for themselves. They live and feed mostly on the streets of cities and villages, or almost permanently encamped at environments that provide a constant supply of food—such as trash dumps. Human-provided food is available in a steady flow at these sites, which results in a stable population of feral dogs.

However, the nutritional quality of this food is much lower than the meat-based diet of wolves. Pariah dogs adjusted to this specific niche with their smallish size—leftovers do not sustain large dogs and are a food source that does not need to be subdued by physical strength. Pariah dogs also seem not to hunt in packs.

Pariah dogs may live in groups of hierarchical organization and show territorial aggression against other groups. They reproduce all year around—mirroring the steady food supply and the supportive climatic conditions. Pariah dog males are constantly pursuing available females, and the females may have two litters yearly. Pariah dog mothers nurture their young only in the first 8–10 weeks,

Above and below
Free-ranging dogs rely on a constant supply of nutrition from human society. They are highly adaptable and usually coexist with humans without causing major problems; otherwise they would not be tolerated.

and there are no helpers (older siblings) or caretaking fathers. Therefore, when the pariah dog puppies are weaned, they immediately face strict competition with the adults for food, and most of them die in the first year.

THE RELATIONSHIP BETWEEN PARIAH DOGS & HUMANS

Importantly, pariah dogs are not "wild," in the sense that they do have some type of relationship with humans. Although they are not socialized like family dogs, the puppies are born in a human-made environment, where traces from the humans (such as olfactory cues, artefacts, visual stimulation) are abundant. Puppies are often adopted by local children, who may offer them to tourists to buy. Adult pariah dogs move confidently around in human settlements and rarely get into conflict with humans. Some citizens routinely feed the local pariah dog groups, with the intention of using them for guarding duties against burglars and other pariah dogs.

CONCERNS ABOUT THE ECOLOGICAL ROLE OF PARIAH DOGS

The phenomenon of domestic species going wild (feralization) is a worrisome tendency, which often has a heavy impact on ecosystems. Cats, ferrets, camels, and other species are documented as burdens on the local fauna or flora in particular parts on the world. Thus, one might assume that the existence of many hundreds of millions of pariah dogs would negatively affect indigenous

DINGOES—THE REAL "WILD DOGS"

It is often (and incorrectly) stated that dingoes represent a separate species of *Canis*. In reality, dingoes are just as human-made as any other dog around the world. They were taken by human settlers to Australia around 4000–5000 BCE and became closely associated with the Aboriginal people. However, many of these dingoes went wild and adjusted successfully to the local challenges of the continent. Nowadays, they are regarded as a pest mostly because they also prey on domestic animals.

species—both as potential predators of prey animals and as competitors of other carnivores. However, because pariah dogs rely mostly on human waste as food, it is less likely that these dogs act as *exploitative competitors* and hunt for the same prey as lions or cheetahs. Their presence may instead make them *interference competitors* to some species living in the wild, such as jackals, badgers, and smaller cats, by harassing them or disrupting their hunts.

CHAPTER 2

Anatomy & Biology

The Dog as a Mammal
& a Carnivore ~⊘

Following tens of thousands of years of domestication and despite showing great variations in shape and function, dogs still show all the characteristics of being a mammal and being a descendant of carnivores.

The "true" mammals emerged in the late Jurassic from some type of reptiles showing several new anatomical and functional features. Among mammals, Eutheria (placental mammals) and Metatheria (marsupials) divided in the Middle Jurassic period and expanded after the great extinction in the Cretaceous–Paleogene, filling in the niches left after the disappearance of the archosaurs. This led to the evolution of the major recent groups of mammals, including carnivores, emerging more than 60 million years ago.

1. Fur evolved with the emergence of early mammals. The fur (or coat) contributes to thermal regulation but also plays a role in camouflage, physical protection, and communication. Dog fur has been modified in many ways during domestication. Sometimes a difference is made between hair and fur. The former is usually characterized by longer growth phases and less shedding. A wide range of colors, not present in wolves, have been selected for in dogs.

7. The sebaceous glands produce sebum, a hydrophobic substance that provides the water resistance and lubrication of the fur and skin. They prevent dehydration through the skin and promote heat insulation. Eccrine sweat glands play a role in thermoregulation, excreting watery sweat that evaporates and cools down the body. Dogs have such glands only on their paw pads and snout, and so rely on panting to regulate body temperature.

6. Mammals are endotherm, which means that they keep their body temperature at a constant level. Dogs' normal body temperature is 101–102.5°F/38.3–39.2°C (human body temperature is 98.6°F/37°C).

2. Mammals evolved a specific middle ear in which three small bones (incus, malleus, and stapes) connect the tympanic membrane to the inner ear. The angular and articular bones that are part of the jaw in reptiles evolved into the incus and malleus of the middle ear. The involvement of these three bones in sound propagation enhanced the hearing range in mammals. Dogs' hearing range is from about 30-40 Hz to 44,000 Hz.

3. During their evolution mammals lost two color-sensitive cones in their retina, becoming dichromatic—sensitive to only red and blue light.

4. Mammals have a distinct jaw structure with the dentary bone carrying the permanent teeth (which are replaced only once from the infant set), which is linked by an articulation to the squamosal bones of the skull. Adult dogs have 42 teeth, 22 on the lower jaw. The milk teeth erupt around day 20 and stay in place until the fifth to sixth month of age depending on the breed.

5. The mammary glands are a special type of sweat gland producing milk, a nourishing excretion rich in fats, sugars, and proteins, used for feeding offspring. These glands are present in both sexes, but become active only during the lactation period after gestation. Their development and functioning are controlled by female sex hormones including estrogen and prolactin. The milk of dogs is especially rich in fats and proteins and contains less sugar than that of humans.

Above There are seven key mammalian traits that distinguish the around 6,000 recognized living species of mammal from other animals.

The Canine Skeleton & Locomotion ~

Like many other predatory mammals, dogs are powerfully built animals, with a skeleton that supports their weight and allows both sprinting and endurance. Although the canine general shape and form has been the result of millions of years of selection, recent breeding strategies have altered the skeleton in various ways. Despite the huge body size differences (which may reach more than a hundredfold between a Chihuahua and a Great Dane), all dogs share roughly the same number of bones—an average of 319—depending on tail length and whether dewclaws are present.

Above right *Both the smallest (2 in./6-7 cm) and the tallest (42 in./110 cm) dogs share roughly 113 more bones than found in humans. Toy breeds' skeletons mature in approximately 6 months, three times faster than those of giant breeds.*

Right *Bones not only enable mobility, but also support and protect the organs of the body, and store minerals. Some of them also produce red and white blood cells, and buffer the blood against excessive pH changes.*

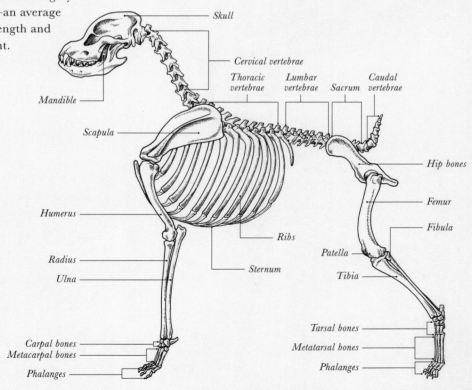

THE DOG'S SKELETAL SYSTEM

- Skull
- Cervical vertebrae
- Thoracic vertebrae
- Lumbar vertebrae
- Sacrum
- Caudal vertebrae
- Mandible
- Scapula
- Hip bones
- Femur
- Fibula
- Humerus
- Ribs
- Patella
- Radius
- Sternum
- Tibia
- Ulna
- Tarsal bones
- Metatarsal bones
- Carpal bones
- Metacarpal bones
- Phalanges
- Phalanges

SKULL

Compared to equally sized wolves, dogs tend to have 20 percent smaller skulls and a more domed forehead. As the extremes, there are long-skulled (dolichocephalic) and short-skulled (brachycephalic) breeds. Skull shortening has led to the reorganization of the brain, which could be an explanation for poor olfaction and better visual acuity in short-skulled breeds.

TEETH

Adult dogs have 42 permanent teeth, including 12 incisors, used for shearing and grooming; 4 long canines, which grasp and tear with great pressure; 16 premolars with sharp edges; and 10 molars, which have a rugged surface used for grinding. Puppies have 12 premolars and do not have molars. Permanent teeth erupt at 3–7 months, and this can be accompanied by drooling and irritability.

HIPS

In a normal hip the rounded head of the femur fits deeply into the acetabulum, a concave indentation of the pelvis. Due to careless breeding, malformations in the hips have become endemic in many breeds. In a dysplastic hip, the head of the femur fits loosely or may be entirely dislocated from the socket.

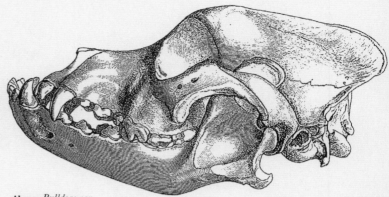

Above *Bulldogs can have a hard time with chewing because of the extreme shape of their jaws, with the upper jaw being shorter than the lower.*

Brachycephaly / Bulldog

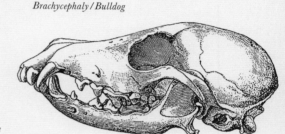

Right *The elongated skull shape is more typical for wild canids. In dogs, some breeds have retained the feature, or even were selected for this narrower head.*

Dolichocephaly / Dachshund

THE TONGUE

The most spectacular muscular organ of a dog is the tongue. It is made up of three major muscles on either side. The tongue is rich with capillaries, helping the dog regulate its body temperature during panting. The world record for the longest dog tongue is 17 in./43 cm and belonged to a boxer.

PAWS

Dogs walk on their toes. The digital and metacarpal pads work as shock absorbers, the carpal pad works like a brake on slippery surfaces. The paw pads' fatty tissue insulates the inner tissues from extreme temperatures and they also contain sweat glands, so they have an important role in thermoregulation. When dogs feel stressed their paws also exude moisture, just like our hands do. Some breeds, Newfoundlands for example, have long toes and webbed feet that make them excellent swimmers.

A dewclaw is an additional digit on the inside of the front legs and occasionally also on the hind legs, positioned analogously to a human thumb. In a standing position, it does not make contact with the ground.

ELBOWS

The elbow is a complex joint, bearing 60 percent of body load. Front leg lameness in young large and giant breed dogs is usually caused by a type of elbow dysplasia, where the three bones of the joint (humerus, radius, and ulna) do not meet properly.

TAIL

Depending on tail length, the number of bones varies between 6 and 23. Dog tails can be straight, sickle, curled, corkscrew, or missing. Tail docking is illegal in many countries.

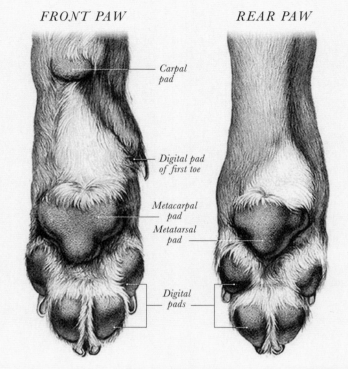

FRONT PAW *REAR PAW*

Carpal pad

Digital pad of first toe

Metacarpal pad

Metatarsal pad

Digital pads

MUSCLES

The muscular system of dogs facilitates movement and provides stability for the joints of the body. There are different types of muscle tissues depending on the job they have to do:

1. Smooth muscles are found in the walls of hollow internal organs, such as the stomach and blood vessels. They control blood pressure by changing the diameter of the blood vessels. These muscles in the walls of the intestines move food along.

2. Skeletal muscles are connected directly to bones in the skeletal system by tendons. As the muscles contract, the tendons pull on the bones. Most dog joints have at least two muscles for each kind of movement—for flexion and extension, for example, or inversion and eversion.

3. Cardiac muscles are present only in the heart, and they share many cellular features with the skeletal muscle. Their main role is to make the blood circulate the body by contracting and relaxing the heart. The resting heart rate is 70 to 160 beats a minute, depending on the size of the dog.

Left *The basenji, an ancient central African breed, has a tightly curled tail, which straightens out for greater balance when running at top speed.*

FORMS OF LOCOMOTION

Breeds vary in their preferred gait. For example, greyhounds do not normally trot—they either walk or gallop, and their foot placement depends on leg length and angulation, too.

A *walking dog* usually has three feet on the ground at the same time. The front feet take the larger share of carrying the body and the rear legs' role is moving the dog forward. Front legs are also more involved in making the dog stop. Typically, a slowly walking dog places its rear foot in the print left by the front foot on the same side to save energy when walking in sand or snow.

Dogs are said to *trot* when the diagonal pairs of legs move synchronously, and generally only two legs at a time touch the ground. Some trotting prints almost appear to be a single line. Dog with longer legs or a shorter backbone have difficulties in trotting because the movement of the rear legs interferes with the movement of the front legs.

A *pacing* dog moves by swinging the front leg and the rear leg on one side while the weight is carried by the other two legs standing on the ground. Pacing, or an ambling gait, is more common in dogs that have difficulties with trotting. Tired, exhausted dogs or dogs that have orthopedic problems may display this type of locomotion.

A *gallop* is employed if the dog needs to move with high speed. During the gallop dogs may have no or only one leg touching the ground. The galloping dog gains speed by decreasing the duration of the stance phase, and increasing the duration of the swing phase in comparison with the walk or trot. Dogs can gallop at two speeds, the slower of which is called a canter. Dogs can maintain a fast gallop for only a limited time.

The Canine Coat & Skin

THE COAT

A dog's coat may be a double or single coat. A double coat or fur coat is made up of a soft, dense undercoat that serves as insulation, and a tougher topcoat, made of stiff hairs to repel water and dirt. Dogs with single coats, such as greyhounds, salukis, or poodles, shed a lot less and need less grooming than dogs with double coats, such as the Australian shepherd dog, cairn terrier, or golden retriever.

Dogs lose old or damaged hair by shedding. This can depend on the season, as many dogs shed their thick winter coat in the spring. However, dogs kept indoors tend to shed all year.

Coat color & behavior

Breeding from the most docile, less stressed dogs led to hormonal and neurochemical changes associated with behavior via metabolic pathways. The genes associated with these changes may also be involved in the physiological control of coat color. Silver foxes selected only for tameness also show alterations in coat color and pattern. Coat color may be linked directly to behavior: Golden English cocker spaniels tend to react more aggressively, while parti-colored individuals of the same breed are more mild-mannered. Extreme depigmentation can correlate with neurological problems and vision and hearing impairments.

Below *There is a greater variety of coat colors, patterns, lengths, and textures found in the dog than in other canids. Coat types were selected to fit the climate and working environment.*

Blue / Weimaraner　　*Bicolor / Border collie*　　*Domino / Afghan hound*　　*Belton / English setter*　　*Harlequin / Great Dane*

Colors & patches

Domestication vastly enlarged the variety of coat colors and patterns. Here is a list of the most frequent patterns. The same color may be referred to differently and the same term may mean different colorations in different breeds.

Belton: Various speckled colors. Also called: flecked, ticked, speckled (English setter, German shorthaired pointer)

Bicolor: Any color coupled with white spotting. Also called: two-color, Irish spotted, flashy, patched, tuxedo, piebald (border collie), Blenheim (for the reddish brown combination in the Cavalier King Charles spaniel)

Blue: Bluish gray (chow chow, Great Dane). Also called ashed.

Brindle: A pattern of alternating stripes of different pigmentation, such as yellow and black ("tiger stripe"), red and black, cream and gray (Great Dane, Boston terrier)

Brown: Depending on the breed and exact shade, this is also called chocolate, liver, mahogany, red, sedge, and dead grass (Labrador retriever, Chesapeake Bay retriever)

Domino: Used to describe the pale overlay covering the top and sides of the body, head and tail, and the outside of

HYPOALLERGENIC COAT

Some dog breeds, such as poodles and bichons, are famous for having "hypoallergenic" coats because they shed very little. In fact, no canine is completely non-allergenic. Not only fur but also dogs' saliva or dander can cause allergic reactions. The reaction to individual dogs may vary greatly among allergic persons, independently of the breed of the dog.

the limbs in Afghan hounds. A similar pattern is called "grizzle" in salukis.

Fawn: Typically refers to a yellow, tan, light brown, or cream dog that has a dark melanistic mask (pug, English mastiff)

Harlequin: Ragged black spots on a white background in the Great Dane

Merle: Marbled coat with darker patches and spots among many white areas (sheltie). Called dapple in dachshunds.

Tricolor: A combination of three clearly defined colors (beagle, smooth collie, sheltie)

Merle / Mudi   *Brown / Labrador*   *Fawn / Pug*   *Tricolor / Beagle*   *Brindle / Greyhound*

THE SKIN

The skin is the largest organ. It guards against dehydration, protects from exposure to the weather, amd helps regulate body temperature through the blood vessels and by muscular action that fluffs the hair and traps heated air next to the body. Dogs do not sweat on areas that are covered with fur but they sweat through the paw pads that make their paws better at gripping the ground. As the first line of defense, skin is a barrier against injury, disease, and damage from the ultraviolet rays of the sun. Pigments in the skin are a natural sunscreen.

Skin layers

The *epidermis* is composed of keratinized cells, which slough off continuously. New cells rise from the basal cell layer. The epidermis contains no blood vessels. In skin infections, the "watchmen" of the immune system—the Langerhans cells—capture foreign proteins (antigens) and process them. If they are overactive, the dog can develop an allergy, which results in itching.

The *dermis* is mostly made up of connective tissue. It harbors many receptors for touch and pain that provide the dog with its sense of the environment. It also contains the hair follicles, glands, and blood vessels.

Below the dermis lies the *hypodermis*, which contains a lot of fat. Fat serves as a protective shock absorber, food storage locker, and insulation for the body.

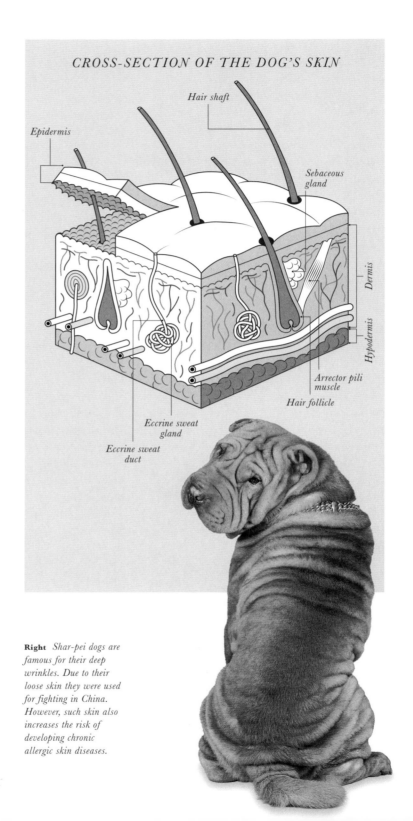

CROSS-SECTION OF THE DOG'S SKIN

Hair shaft

Epidermis

Sebaceous gland

Dermis

Hypodermis

Arrector pili muscle

Hair follicle

Eccrine sweat gland

Eccrine sweat duct

Right *Shar-pei dogs are famous for their deep wrinkles. Due to their loose skin they were used for fighting in China. However, such skin also increases the risk of developing chronic allergic skin diseases.*

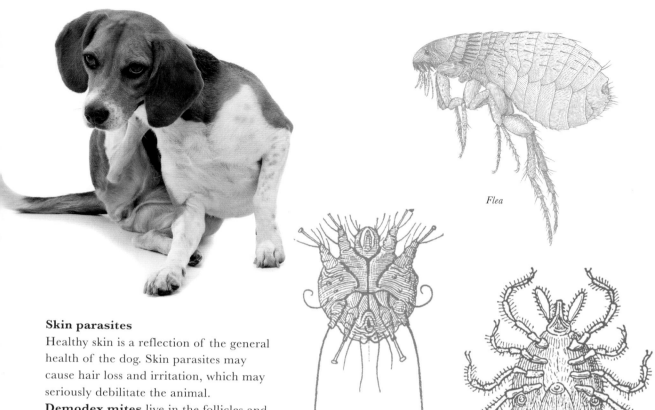

Flea

Sarcoptes mite

Tick

Skin parasites

Healthy skin is a reflection of the general health of the dog. Skin parasites may cause hair loss and irritation, which may seriously debilitate the animal.

Demodex mites live in the follicles and in predisposed dogs cause red mange (demodicosis), which is not contagious to humans.

Sarcoptes mites burrow into the skin and infest not only dogs but cats, pigs, and even humans. Affected dogs need to be isolated from others, and should receive veterinary treatment. Places they have used must be cleaned.

Fleas are small flightless insects and are vectors for tapeworms. Flea saliva may also cause allergies. Fleas normally concentrate in the skin of the head, neck, and tail. Fleas' feces are visible on the skin as tiny black spots the size of black pepper. They can be spotted and caught by using a flea comb. Although dog fleas usually feed on the blood of dogs and cats, sometimes they bite humans. Before using a flea-control product, it is wise to consult a veterinarian.

Ticks are eight-legged arthropods and can transfer Lyme disease, babesiosis, and other infections through their bites. Ticks climb or drop onto the dog's coat when the animal brushes past the area they are sitting in waiting for their prey. The best preventive is to use an insect repellent and examine the dog after a walk and remove any ticks. A tick feels like a small bump on the dog's skin. Rapid removal (using a tick-remover tool or tweezers) lessens the risk of disease. The symptoms of tick-born disease include fatigue, appetite loss, fever, lameness, and blood in urine.

Above left *Dogs scratch to get rid of various external parasites. Nevertheless, owners should also check regularly for the presence of ticks, fleas, and mites because they can carry dangerous diseases.*

Above *External parasites feed on body tissues. The wounds and skin irritation they produce result in discomfort and can cause serious skin problems.*

Eating Like a Dog

It seems to be a paradox that, while dogs have a straightforward ancestry among predators that feed on predominantly meat, hundreds of millions of pariah dogs around the world are seemingly thriving on the lowest quality of garbage. Why is it so complicated to find the ideal diet for our dogs? The cause of the dilemma is most probably a cultural one: Companion dogs are so embedded in the world of humans that their nutrition generates debates as heated as those surrounding human nutrition.

ARE DOGS PREDATORS?

Dogs belong to the carnivores, having the typical canine jaw and dentition indicating the feeding behavior of a larger predator. Dogs can inflict serious wounds by their fangs, and hold fast to their prey despite vigorous escape attempts. Another inherited trait of the pack-hunting wolflike ancestor in dogs is the capacity (and preference) for consuming large, infrequent meals—coupled with a tendency to show harsh competition around food with other dogs.

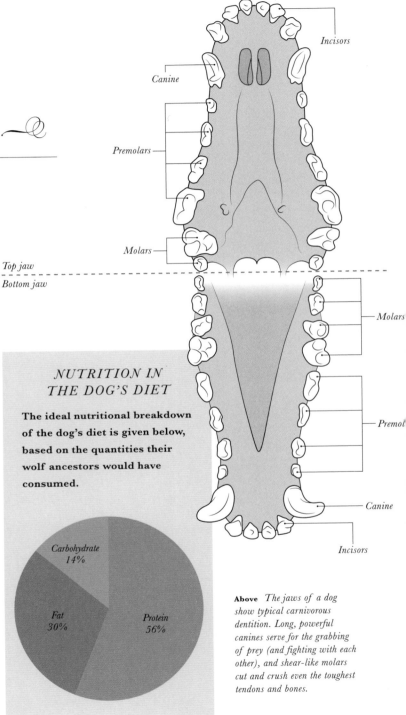

Incisors

Canine

Premolars

Molars

Top jaw
Bottom jaw

Molars

Premol

Canine

Incisors

NUTRITION IN THE DOG'S DIET

The ideal nutritional breakdown of the dog's diet is given below, based on the quantities their wolf ancestors would have consumed.

Carbohydrate 14%

Fat 30%

Protein 56%

Above *The jaws of a dog show typical carnivorous dentition. Long, powerful canines serve for the grabbing of prey (and fighting with each other), and shear-like molars cut and crush even the toughest tendons and bones.*

FOODS DOGS MUST NOT EAT

There are several types of foods that dogs must not eat. These include avocado, grapes, raisins, chocolate, coffee, macadamia nuts, and any foods that contain xylitol (such as chewing gum).

In addition, there is a wide range of human food that may make the dog ill, for example sugar, citrus oil, and mushrooms.

Raw chicken

Cereal based (wet)

Cooked chicken

Cereal based (dry)

Above *There are multiple options for the diet of our canine companions, especially in wealthier countries, where nutritional value, convenience, and fashion can play equally important roles.*

VARIABILITY IN FEEDING

Pariah dogs that do not get direct nourishment provided by people feed mainly on meals of animal origin. In Brazil, local feral dogs eat mostly whatever small vertebrates they can find, such as lizards and mice, and insects. In Africa, feral dogs compete with indigenous scavengers for carrion.

Laboratory tests have shown that companion dogs prefer meat over cereal-based diets; their favorite types of meat are pork, beef, and lamb over chicken, horse, and liver. Dogs have also been found to prefer cooked meat over raw.

A dog that weighs 110 lb./50 kg needs 2,400 kcal daily, which is comparable to the requirements of a 35-year-old, 155-lb./70-kg man. The composition of a dog's diet should be protein-biased (daily approximately $^{1}/_{8}$ oz. protein/5 lb. of bodyweight (1.5–2.5 g protein/kg of bodyweight); smaller dogs require relatively larger amounts of protein. Proteins of animal origin are preferred, and the ratio of fat to protein should be around 1:3–1:4.

However, there is major disagreement about the source of nutrition. While many experts say that factory-made dog foods assure the scientifically tested "ideal" ratio of macro- and micronutrients and vitamins as well as the adherence to rigorous hygiene standards, other experts advocate natural (nonprocessed) ingredients, often in their raw form.

EATING & HEALTH

Food intolerance and allergies threaten the welfare of millions of dogs (and their owners' as well). Rather shockingly, dogs can become allergic to certain "natural" compounds of their diet, such as different kinds of meat. Not surprisingly, there are multiple suspects for this sad scenario, including intestinal infections at a young age; artificial chemicals either in the food or in the environment; or inbreeding depression. Infections alone could not cause high-frequency allergies in a species that has evolved to digest less-than-hygienic food items, and this enhances the likelihood of genetic predisposition for allergies in modern-age dogs.

Canine Physiology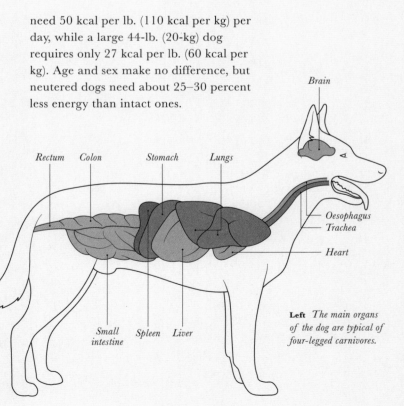

The main role of physiological processes is to keep the body in balance (homeostasis) so that the dog can react to any changes in its inner or outer environment. The proper working of all organ systems prevents the dog from developing any pathological condition.

HEART & CIRCULATION

One of the main organs is the heart, whose role is to keep blood moving round the body. Only by continuous circulation of the blood can all parts of the body get access to energy-rich molecules and oxygen. At the same time, blood dispatches waste products and carbon dioxide. Dogs have approximately 3 fl. oz. blood/2.2 lb. (80–90 ml blood/kg), which is pumped around by the heart beating between 70 and 160 beats per minute.

THE DOG'S ENERGY NEEDS

The maintenance energy requirement (MER) of an adult dog is the minimum energy per day that a dog needs at typical levels of activity. This varies in companion dogs, depending on living conditions, activity, and breed. Small dogs use relatively more energy than large dogs: a 4.4-lb. (2-kg) dog may

GOOD STRESS

It may seem contradictory but to be healthy dogs need to be exposed to stress. "Good stress" in this case means that the physiological state is combined with the feeling of positive emotions, such as playing or being in close social contact. Some limited experience of negative stress may also be important because this is the way that dogs can learn to cope with the challenges that may occur in their lives.

need 50 kcal per lb. (110 kcal per kg) per day, while a large 44-lb. (20-kg) dog requires only 27 kcal per lb. (60 kcal per kg). Age and sex make no difference, but neutered dogs need about 25–30 percent less energy than intact ones.

Brain

Rectum　*Colon*　　　*Stomach*　　*Lungs*

Oesophagus
Trachea

Heart

Small intestine　*Spleen*　*Liver*

Left *The main organs of the dog are typical of four-legged carnivores.*

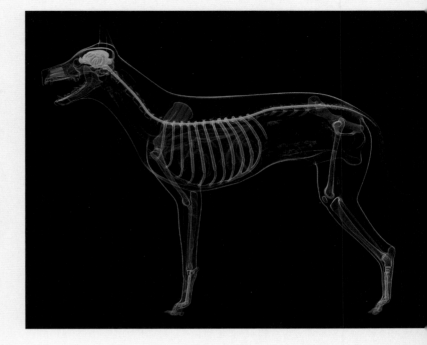

MAINTAINING HOMEOSTASIS

Many factors affect the physiological control processes—not just hunger and thirst, but also external stimuli, such as decreasing ambient temperature or being frightened by another dog. These effects can be generally summarized as some sort of stress. Stress activates specific neural structures (the sympathetic nervous system) and hormones (such as cortisol) that help the body to maintain its balance. For example, if a dog starts to play with its owner its body needs more energy, so the activation of the sympathetic nervous system makes the heart work faster, and more energy-rich molecules stream from the depots of the body to the muscles and brain. If there is no need for extra energy, the parasympathetic nervous system controls the physiological processes.

CORTISOL & STRESS

Cortisol is often referred to (wrongly) as a "stress hormone" because it is assumed that elevated levels of cortisol mean a negative inner state. Indeed, during storms dogs with thunderstorm phobia show elevated levels of this hormone, while aggressive interaction between dogs or between dogs and humans also leads to high cortisol secretion.

However, cortisol concentration also increases if dogs are playing with their owner or with one other. Thus, cortisol is more an indicator of an imbalanced physiological state in either a "positive" or "negative" direction.

VITAL STATISTICS OF DOGS & HUMANS

HUMAN	DOG
Body temperature	
101–102.5° Fahrenheit (38.3–39.2° Celsius)	97–99° Fahrenheit (36.1–37.2° Celsius)
Respiration rate	
10–12 breaths per minute	10–35 breaths per minute
Blood pressure	
90–120 mmHg (systolic) 60–85 mmHg (diastolic)	130–180 mmHg (systolic) 60–95 mmHg (diastolic)
Heart rate	
50–90 beats per minute	140–180 beats per minute (smaller breeds have higher bpm)

How Dogs See

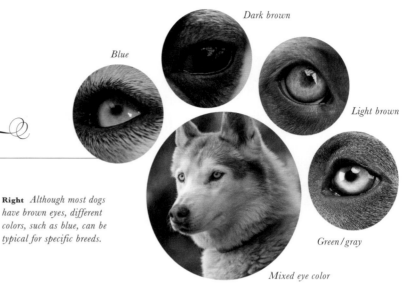

Canines seem to possess a typical mammalian visual system that is particularly well adapted to function under low light conditions. Dogs rely more on their visual abilities than one would expect, because in the human (social) environment visual information is of more significance. Artificial selection, however, has led to an enormous variation in dogs' head shape, eye position, and the structure of the retina. Thus, dog breeds may differ markedly in their visual abilities.

Right *Although most dogs have brown eyes, different colors, such as blue, can be typical for specific breeds.*

WHAT DOGS SEE

Humans may assume that dogs share their visual perspective, seeing the world in a similar way. This is, however, not true for several reasons. Firstly, being taller, humans have a much better overview of a terrain than dogs. On the other hand, dogs' visual angle is wider because their eyes are placed more laterally on the head, and thus they do not need to move their head when scanning the landscape.

CHANGES IN HEAD SHAPE AFFECTED THE ABILITY TO FOCUS

The ability to focus on and attend a target object depends partly on directing the incoming image of the object onto the fovea, which is the point of sharp vision on the retina. Most dogs with longish heads (dolichocephaly) do not have a fovea, and the area of sharp vision has the shape of a horizontal stripe. However, selection for dogs with short heads (brachycephaly) has affected the structure of the retina, so many breeds such as pugs have acquired a fovea-like location in the eye.

Dolichocephaly

Brachycephaly

COMPARISON OF DOG & HUMAN COLOUR RANGES

Trichromatic

Dichromatic

THE DOG'S FIELD OF VISION

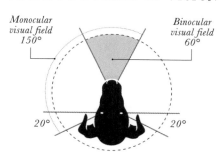

Monocular
visual field
150°

Binocular
visual field
60°

20°   20°

VISUAL PROCESSING ABILITIES

Sensitivity to brightness is very important for a predator that hunts under low light. In general, dogs are half as sensitive to shades of gray as humans, but they are more sensitive to light in general. Thanks to a special layer (the *tapetum lucidum*) at the back of the dog's eye that acts as a mirror, incoming light is reflected back into the eye. This makes the attended scene brighter for dogs than for humans, who lack this feature.

Acuity refers to being able to separate visually two objects from a distance. The visual acuity of humans is especially good—we can discern very small gaps between two vertical stripes from far away. Using the same stimulus arrangement, it has been shown that, if humans can tell that two parallel stripes are separate from a distance of about 25 yd./22.5 m, then dogs need to approach the stimulus at 20 ft./6 m to see the same pattern.

Being descendants of predators, dogs have an enhanced sensitivity to detect motion. They may notice even very tiny movements from a relatively long distance that escape human observation. This feature helps canines in general to locate small prey, and this may also explain why

many dogs have problems finding nonmoving objects in the grass.

The dog's retina consists of two types of specific receptor cells (cones) that are most sensitive to colors that appear to humans as blue-violet and greenish-yellow. Thus, dogs have a dichromatic visual system, while humans are trichromatic (shown in the diagram above). Dogs can easily discriminate yellowish versus blueish objects but do not see red-like colors, instead perceiving them as some sort of gray.

Above *The dog's visual field is larger than that of humans. The coverage of the dog's binocular field is about half the typical human value, which explains why dogs have relatively poor depth vision.*

Below *The dog's eye has the same components as the human eye, but the dog's cornea and lens are more rounded than their human counterpart.*

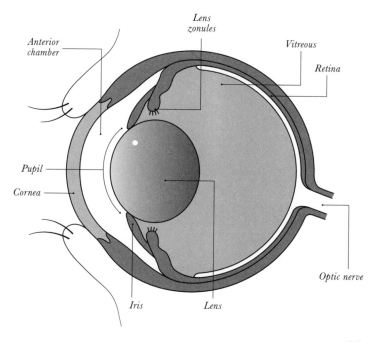

Anterior
chamber

Lens
zonules

Vitreous

Retina

Pupil

Cornea

Iris   Lens

Optic nerve

How Dogs Hear

Humans may have decided that dogs would make good guards when they noticed that they are very sensitive to noises, especially at night. An alert dog, vocalizing upon approaching strangers, could be a useful aid against predators as well as human intruders.

THE DOG EAR ACTS LIKE AN ANTENNA

Good hearing is a crucial ability for any predator, especially for those that hunt under low light. The sound of prey can travel a long way and is less obstructed by trees, bushes, and rocks than the sight of it is, so a predator with fine hearing skills can easily detect the presence, distance, and direction of the prey. In canines, this ability is markedly enhanced by a moveable outer ear. The rotation of the ear helps the dog to localize the sound source in space. Many dog breeds have floppy ears as a result of artificial selection that may impair this ability.

Ear movement is also a good visual indicator for an observer to know what the dog is listening to. Dogs that look forward attentively at their owners but have their ears turned back may be concentrating on the sounds they are hearing rather than watching their human companion.

DOGS EXCEL AT HIGH SOUND FREQUENCIES

It is well known that dogs' hearing range is much broader than that of humans. While dogs and humans are equally equipped to hear low-frequency (deep) sounds (~30–40 Hz), dogs' upper hearing of high-frequency sounds (~44,000 Hz) goes much beyond the human maximum (~18,000–20,000 Hz). This means that dogs hear rather well in the ultrasonic range, and thus probably hear the high-frequency vocalizations produced by small rodents such as rats or mice. The bias toward high frequencies may be advantageous because small prey often produce such sounds when moving

Below left *Hanging ears might have emerged as a result of domestication, and were probably not selected for.*

Below right *Erected ears can help with localizing prey or companions from a distance.*

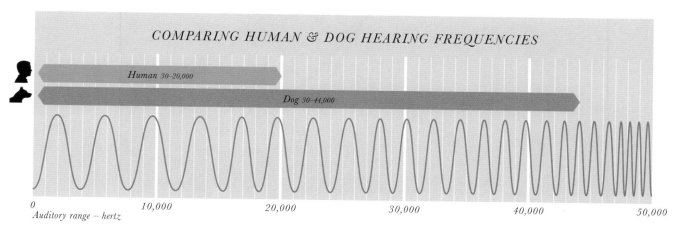

COMPARING HUMAN & DOG HEARING FREQUENCIES

Human 30–20,000

Dog 30–44,000

0
Auditory range – hertz   10,000   20,000   30,000   40,000   50,000

around. This also means that dogs could be more easily disturbed by ultrasonic noises that are produced by machines or household equipment, but which are not heard by humans. One practical advantage of dogs' high hearing range is that they can be trained to return upon hearing an ultrasonic whistle that cannot be heard by humans standing by.

LOCALIZATION OF THE SOUND SOURCE

Directional hearing can be tested by finding out whether the subject is able to localize exactly a sound source from a long distance. Although not much data is available, humans excel at such tasks because they are able to tell apart locations that deviate 1° in space, while dogs can discriminate objects at angles of 8°. There is evidence that the outer ear helps in localizing sound sources, but there is no research to show whether dogs with erect or floppy ears have differing abilities in this respect.

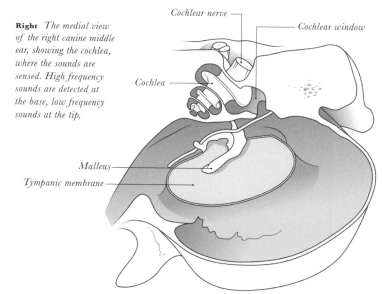

Right *The medial view of the right canine middle ear, showing the cochlea, where the sounds are sensed. High frequency sounds are detected at the base, low frequency sounds at the tip.*

Cochlear nerve
Cochlear window
Cochlea
Malleus
Tympanic membrane

SURGICAL MODIFICATION OF OUTER EARS MUST BE AVOIDED

There are some breeds in which surgical modifications of the outer ear are necessary to obtain the "official" look. Although ear cropping is banned in many countries around the world, including Australia, New Zealand, and much of Europe, many dogs are still put through this unnecessary procedure. Surgical intervention should be initiated only on the basis of medical advice.

How Dogs Smell & Taste

According to some theories dogs may have been domesticated because of their superior olfactory capacity. Humans are not particularly talented in gaining olfactory information, so working with dogs could have magnified humans' success in hunting. While we do not know whether this is a plausible scenario, dogs' skill at smelling is often described as extraordinary. It is no wonder that dogs have been deployed as helpers in the hunt, and more recently utilized as search and rescue dogs or detectors for the police and security services.

ENHANCED OLFACTORY ABILITY IN DOGS

Dogs' enhanced olfactory ability relies mainly on two factors. Dogs have, just as their wolflike ancestor did, a relatively large nose cavity that contains a troglodytic structure consisting of a complex system of small bony walls (ethmoturbinates). This large folded surface is covered by the olfactory epithelium, which is the sensing organ for molecules reaching the nose cavity. The total surface of the olfactory epithelium in dogs may reach 23 in.^{2}/150 cm^2, while in humans it is almost 1 in.^{2}/5 cm^2. This means that there is a much higher chance in dogs of a few molecules of a substance reaching the large olfactory epithelium compared to humans.

Left *The ability to detect chemical compounds at low concentration is especially important for hunting dogs, such as the Hungarian vizsla.*

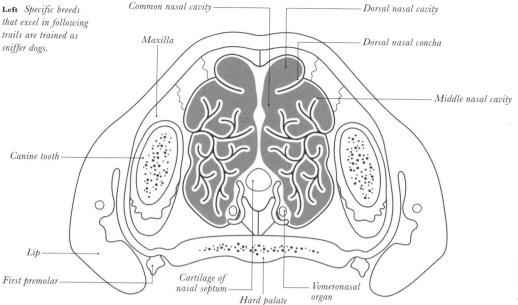

Common nasal cavity

Dorsal nasal cavity

Maxilla

Dorsal nasal concha

Middle nasal cavity

Canine tooth

Lip

First premolar

Cartilage of nasal septum

Hard palate

Vomeronasal organ

Dogs have more active olfactory receptor genes than humans (30 percent more), which determine those protein molecules that sit in the cell membrane of the olfactory neurons and are the receptors for the odorous compounds carried in the air. Therefore, when breathing in a possible odorous molecule, dogs have a higher chance of having a matching receptor sitting in the cell membrane of the olfactory epithelium that is activated by that smell.

Left *Cross-section of the middle of the dog's nose, with the nasal cavity in gray. Dogs have a larger nasal cavity than humans.*

Below *The rhinarium (the tip of the nose) is usually cold and moist. It is also sensitive to wind direction, thus it helps the dog to sniff in the right direction.*

DOGS DETECTING CANCER

In recent years many dogs have been trained to detect the presence of cancer in humans. It was assumed that dogs might be able to recognize minute differences in the chemical composition of healthy and sick tissues. Many studies reported that dogs were able to indicate urine samples from people with prostate cancer, and they also picked out humans who had lung cancer.

While it is unlikely that these dogs can provide a diagnosis for patients, investigating their detection performance may give a clue to researchers seeking to find which molecules could be used as biomarkers for a specific illness.

THE ACT OF SMELLING

Smelling is an active process and dogs can inhale air at a rapid rate, approximately 4–7 Hz (sniffs/second). This ensures that odorous molecules also reach the deeper parts of the nose cavity, and about 15 percent of them stay there for the next round of smelling. This increases the concentration of the chemical in the nose and also provides more time for analysis. Sniffing behavior also changes during a search. Dogs smell very effectively at the beginning when they need to locate the start of the trail. After finding the start, dogs may wander along in a more relaxed way and sniff less frequently.

OLFACTORY ACUITY

Several experiments have established that, in the case of some chemicals, dogs have a much higher sensibility than humans. Depending on the molecule, this difference could be 3–10 times higher, or even 10,000 (in the case of n-amyl acetate). In most cases, dogs need to be trained to recognize specific odors (such as narcotics or components of explosives). After training, skilled dogs are able to detect these substances at minute concentrations.

Although differences between breeds are expected, they have not been documented well. There is some evidence that the spontaneous performance of scent dogs (such as the beagle) and wolves is higher than that of non-scent dogs (such as the Afghan hound) and short-nosed ones (such as the boxer).

Above *The German shepherd is a breed that is highly suited to tracking work.*

RECOGNITION & MATCHING OF ODORS

Odor detection may be important but in other tasks dogs need to be able to match odors. A well-trained dog can achieve a near 100 percent success rate in tasks involving odor recognition and discrimination. A trained dog may search for a path based on a few sniffs from an object. This skill is very useful if there is a need to search for a lost person. Dogs can also boost police investigations if they can match an odor trace found at a crime scene with an odor on the clothes or other objects of a suspect. Dogs can even discriminate very reliably between monozygotic twins.

VOMERONASAL ORGAN

This specific organ for smelling sits in the roof of the mouth (a little behind the upper incisor teeth) and has a duct that opens into the nasal cavity. The function of the vomeronasal organ as an additional smelling organ is still a mystery, although it is generally assumed that it may be especially sensitive for pheromones associated with sexual activity. Based on observation of other mammals, male dogs with an impaired vomeronasal organ may not show typical mounting activity while females with the same impairment may not accept male dogs' courtship.

TASTE PERCEPTION

Because dogs are not obligate predators, their taste receptors are more like those of humans. Receptors that are sensitive for one of the five basic tastes (sweet, bitter, sour, salty, umami) sit in a specific taste bud at one specific region of the tongue. For example, sweet is sensed at the tip of the tongue.

There are many different receptor proteins that contribute to taste sensing. Dogs have 15 genes that code bitter-sensing receptors at the back of the tongue. Dogs' preference for sweet tastes can be dangerous because they may encounter sweet foods in the household that are poisonous to them (such as chocolate and xylitol).

EARLY PREFERENCE LEARNING

Dog puppies show a preference for substances that they encountered as embryos during pregnancy. For example, puppies that are a few weeks old prefer anise-scented food if their mother was fed with anise during pregnancy and lactation. Such food preferences may help inexperienced offspring to eat safe food after weaning.

Below *As puppies are fed by their mother, they pick up the taste for foods that she has been eating.*

The Genome ~Q

Geneticists have reached the conclusion that dogs have approximately 18,500 genes, which is on the same magnitude as the number of human genes (19,500). Soon after the human genome project was finished, the full genome of a boxer named Tasha was sequenced in 2004.

CHROMOSOMES

Dog genes are packaged into 38 pairs of body chromosomes and one pair of sex chromosomes (humans have 23 pairs in total). One of each chromosome pair is inherited from each parent at the time of conception. Genetic traits linked to the sex are passed on through the sex chromosomes. Male dogs have one X and one Y sex chromosome, while females have two Xs.

Despite the large evolutionary distance between them, dogs and humans share many parts of the chromosomes, and this similarity is higher between humans and dogs than between humans and mice. However, some of these matching sections are found on different chromosomes in dogs and humans. Dogs and humans may also share many genes that have the same effect. Thus, in both species any mutations may lead to similar malfunctioning.

THE EFFECTS OF GENES ON THE DOG'S SHAPE

It has been known for a long time that in humans the gene named insulin-like growth factor 1 (IGF-1) is involved in determining growth and a person's final size and weight. Recent research has found that this gene plays the same regulatory role in dogs. Variations of this gene (alleles) in dogs determine whether the dog remains small (miniature schnauzer) or big (giant schnauzer). In some large breeds 40 percent of the size is controlled by specific IGF-1.

In short-legged dogs, such as the basset hound and dachshund, the long leg bones stop growing early. This is caused by a partly dominant mutation in the fibroblast growth factor 4 gene (FGF4) that is responsible for limb development.

Above *The basset hound is deliberately bred to have chondrodysplasia, which causes stunted leg growth. The condition is due to an extra copy of the gene involved in growth.*

Below *A variation of the IGF-1 gene is found in all small breeds, but is almost entirely absent from large breeds and gray wolves. This gene plays an important role in determining size in dogs, such as in the miniature, standard, and giant schnauzers.*

Below *Dog puppies in a litter are usually fraternal twins, but there can be identical dog twins too.*

GENES MAY BE ASSOCIATED WITH CHANGES IN DIET

Living in the anthropogenic environment, dogs in general consume much more starch than their wild relatives. It has been shown that dogs have more copies of the gene (AMY2B) coding the alpha-amylase enzyme than the wolf. The role of this enzyme is to break down the large starch molecules into smaller units (oligosaccharides), and more copies of this gene ensure that more enzymes are produced, and the process of digestion is more effective.

Above *Dogs possess genes for digesting starches, making them better than wolves at splitting starches into sugars with the amylase enzyme.*

GENETIC FACTORS

The new tools geneticists have at their disposal have increased the hope that we can reveal the genetic factors that play a role in both canine and human diseases, as most human diseases have canine counterparts. DNA testing also allows breeders to avoid mating disease carriers.

Inheritance of Traits

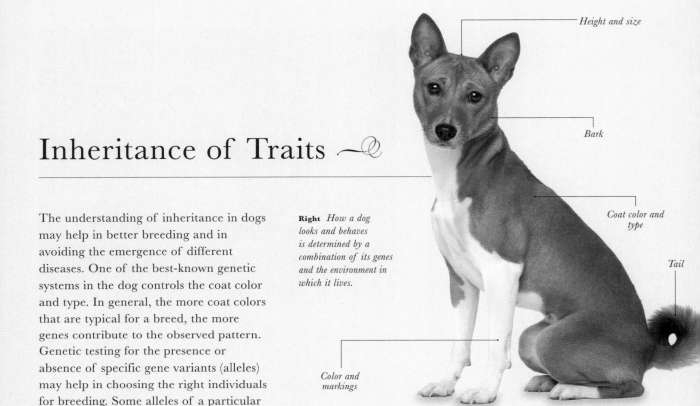

Height and size

Bark

Coat color and type

Tail

Color and markings

Right *How a dog looks and behaves is determined by a combination of its genes and the environment in which it lives.*

The understanding of inheritance in dogs may help in better breeding and in avoiding the emergence of different diseases. One of the best-known genetic systems in the dog controls the coat color and type. In general, the more coat colors that are typical for a breed, the more genes contribute to the observed pattern. Genetic testing for the presence or absence of specific gene variants (alleles) may help in choosing the right individuals for breeding. Some alleles of a particular gene could also have a negative effect on the dog's welfare.

INHERITANCE OF BEHAVIOR

Behavior traits are polygenic, so most genes involved have only a limited effect. The effect of these genes usually depends on the environment in which the dog lives (gene–environment interaction), making the effect of genes on the dog's behavior much harder to detect. Genetic effects on behavior can also be estimated if distantly related breeds are crossed. For example, when researchers crossed barkless basenjis with relatively barky cocker spaniels, dogs in the first generation barked more like dogs from the latter breed.

RELATIONSHIP BETWEEN GENES & BEHAVIOR

The dopamine receptor (protein molecule sitting in the membrane of brain cells) is activated by the dopamine (neurotransmitter). One of the first gene–behavior association studies in humans found that alleles of the receptor gene explain differences in novelty-seeking. Researchers hypothesized that the same gene may affect dogs' behavior in a similar way. Indeed, they found that dogs carrying one specific allele type show increased activity-impulsivity.

GENES & COAT COLORS

There are eight known canine genes associated with coat colors. Each has at least two alleles. These genes control the distribution of pigments in individual hairs, the production of eumelanin (black) and pheomelanin (yellow) pigments, and the intensity of pigmentation.

Agouti: This gene determines whether single hairs are banded (agouti) or of a solid color. Agouti is a common wild type color, and is associated with fearfulness in foxes and rats. In the dog, there are five suspected alleles that control production of the agouti signal peptide (ASIP).

Brown: There are at least two known alleles of this gene that control the production of the black eumelanin pigment. The B (brown) allele is dominant to b (not brown).

Dilute: Different alleles of this gene determine the intensity of pigmentation. The D allele (not diluted) is dominant to the d allele. If a dog possesses two ds, black becomes gray or blue, and brown becomes light tan or "Isabella."

Extension: Five known alleles determine whether a dog expresses the eumelanin pigment in its coat. The presence of the E, Em, Eg, or Eh alleles results in a black or brown coat by allowing eumelanin production throughout the body; the lack of eumelanin makes the coat red or yellow when only the recessive e allele is present (ee). Em causes a dark mask on the head region. The Eg (grizzle) allele modifies the pattern of tan markings created by the agouti (in sight-hound breeds). The Eh allele has been identified in cocker spaniels and creates pheomelanistic markings in black dogs, resulting in a sable-like pattern.

Dominant black: Three known alleles determine the coloring pattern. If the dog possesses at least one dominant allele (KB), then it eliminates the effects of the agouti (ASIP) gene. With other alleles, brindle or agouti may appear. Fully black-colored coats appeared in the dog first, and black wolves acquired their coloration from wolf–dog hybridization.

Merle: Two alleles control the merle pattern (patches of sporadic colored and dilute hairs). The merle (M) allele is dominant to the non-merle (m) allele. The phenotype caused by M is visible only in dogs that are not ee (see above). The presence of the M allele is linked to auditory and ophthalmologic abnormalities, thus mating of two merle-colored dogs is not allowed.

Harlequin: Harlequins can have only one dominant H allele because two H alleles are lethal during early development, and these embryos are absorbed in the womb. For a dog to have the harlequin pattern at least one M allele (merle) must also be present.

Spotting: There are either two or four alleles that determine the degree and distribution of spotting of a dog's coat.

Above *Black wolves owe their coloration to a mutation that first occurred in dogs. Later this gene was carried over to wolves through dog-wolf hybridization.*

COAT TYPES

Long hair: Not all genes causing long hair have been located yet, but mutation of one gene makes the hair grow longer in many dog breeds. Many short-haired breeds may have long-haired puppies occasionally because a long-hair variant can be carried down many generations without being expressed.

Hungarian puli

The thickness of cords depends on the undercoat to outer-coat balance

Thick coats protect from injury

Afghan hound

Long silky coat

Dark overlay

Thick hair for protection against cold conditions

Curly hair: Its appearance depends on which other coat type genes it is combined with. The moplike coat of the Hungarian puli and komondor falls into heavy cords and protects them from both attacking enemies and extreme weather conditions.

Wire hair: The allele causing wire hair is dominant. But another gene (I) may affect its expression, which results in a long, soft coat.

Wirehaired pointer

Right The Mexican hairless dog is an ancient breed, believed to have originated more than 3,000 years ago.

Hairlessness: There are four hairless breeds: the Mexican hairless (xoloitzcuintli), Peruvian Inca orchid, Chinese crested dog, and American hairless terrier. Not all individuals are hairless in these breeds. The first three breeds have dominant hairlessness, meaning that they need one dominant H allele of the hairless gene to be hairless. In fact, all hairless dogs have only one H allele, because embryos that inherit two Hs will be absorbed in the womb.

Both types of hairlessness can have associated health problems, mainly allergies. Such dogs are also likely to get sunburn and skin cancer. Dominant hairlessness may also cause tooth problems.

AN EXAMPLE FOR SIMPLE INHERITANCE: THE HAIRLESS GENE

The possible outcomes of a breeding can be estimated by using a Punnett square (see below). The hairless gene has two alleles, one of which (H) is dominant to the other (h) (recessive allele). Each parent has a 50 percent chance of passing on one or the other allele (individuals always have two alleles of each gene). The dominant allele makes the hair disappear. However, two copies of these alleles are lethal, so hairless individuals always carry one H and one h allele (heterozygotes).

Hairless torso

Hairy feet and head

Right Although it is a hairless breed, the Chinese crested dog has tufts of hair on its paws, tail, and head.

POSSIBLE OUTCOMES OF THE HAIRLESS GENE

Hairless parent alleles

	H	h
H	HH *lethal*	Hh *hairless*
h	hH *hairless*	hh *powder puff*

CHAPTER 3

Behavior & Society

Social Behavior in Canines ~⟨

Being highly social and living in families is a special feature of canines. The basic theme is the same—differences are only quantitative. Family dogs have inherited most social traits present in their wild relatives, but they also need to learn about the peculiarities of social interaction. This is especially important if many dogs are living together in a human family.

THE FAMILY ORGANIZATION

Typically, two or three generations live together in a wolf pack, while groups of jackals and coyotes are usually smaller. The actual organization depends on many factors, and in wolves it is not infrequent for such family packs to join together and form even larger packs of 20–30 individuals. The genetic relationship among the members ensures that pack life is usually peaceful because its success depends both on the parents and the survival of the offspring. Thus, the oldest male that is the father of the younger pack mates is closer to a leader who has the most experience and takes the most decisions. But in the end his interests are likely to concur with those of the family.

When wolves reach two or three years of age, they leave the pack to establish a new family. Given that a specific area is covered by territories of other wolves, this task requires courage and experience. It is not surprising that only a few wolves make it. This is one feature that is not typically present in family dogs because most of them prefer to stay with their human family. Free-ranging dogs disperse at various ages, but they are also more easily accepted by other packs.

Below *Like all jackal species, black-backed jackals live in small family units.*

MALE & FEMALE STRATEGIES

Social interactions in the pack increase as the leader female, the mother, is getting in heat. Field observations suggest that most courtship activity is confined to the breeding pair. During the mating season the father tries to prevent any sexual interactions between the mature males and the mother, and in general is aggressive toward all other males. In contrast, the leading female is more determined to show a low level of intrasexual aggression toward all other females during the whole year.

In free-ranging dogs the wide variety of mating systems does not support such sex-specific roles, and family dogs are also not much drawn into such strategic behaviors in relation to mate finding.

RECONCILIATION

For a canine social group, cohesion is very important. While fights in some cases are unavoidable, making peace between former adversaries is equally paramount. Observational studies both in wolves and in family groups have shown that such reconciliation after aggressive interactions does occur. After the combat either the loser or winner will indicate their preference to stay in the vicinity of the other, and they will engage in body contact and social licking. Reconciliation ensures that the pack members retain their willingness to collaborate, both when they need to defend their territory and when hunting.

HUNTING PACKS

Hunting is a central activity in all canines, but the most complex hunts involving large numbers of individuals have been observed mainly in wolves living in the far north of Canada and Alaska. It is assumed that the typical family size of wolves is also determined by the size of their prey. Wolves live in larger packs if they have to hunt elk or muskoxen, but will hunt alone if their prey is smaller.

Hunting does not consist merely of locating and chasing prey. Wolves need to know their sometimes vast territories very well—where and when prey is moving— and to be able to organize hunts over a range of 12–40 miles (20–65 km). Wolves have been observed to make short cuts or even ambush for a surprise attack. Free-ranging dogs rarely hunt in groups, for simpler tactics suffice to find food near human settlements.

Below *Aggressive interactions may take place within a family pack but they are usually followed by some form of reconciliation.*

FOOD SHARING

Dividing up the prize of the hunt is a common issue in the life of a pack. Generally, it is the leader that is responsible for this task. If the prey is very large (such as an adult moose), then disputes are rare. Quarrels are more intense if a muskox calf has to be shared. Each wolf has an "ownership zone": Even leaders will respect this zone if another wolf already has a piece of meat in its mouth. Food sharing is more intense during whelping. Both parents and all brothers and sisters may share their meal with the young puppies, either by regurgitating or by carrying pieces of meat home from the hunt.

This behavioral trait is also present in dogs, but they need to be socialized to be tolerant to humans who may interfere with their eating, or to respect the food that others possess.

INTER-PACK RELATIONSHIP

In contrast to the relatively peaceful family life, there seems to be no mercy when it comes to inter-pack conflicts. This is the case when wild canines encounter a pack of the same species or of another species. Wolves and coyotes are competitors in North America, just as wolves and jackals are in southern parts of Eurasia. Lone individuals are likely to be chased away from the territory, or often get killed.

All wild canines learn when they are young who belongs to their family, and who is a stranger. This learning is repeated when they establish their own family. In contrast, family dogs are expected to readily accept strange dogs,

and even to show affiliative behavior toward them. Most dogs can deal with such situations but they need to be socialized in order to develop this level of social competence.

MULTIPLE-DOG HOUSEHOLDS

It is popular to have more than one dog in one household. The way these canine groups organize themselves depends on several factors. Somewhat paradoxically, in these cases it may be more legitimate to speak about a "dominant" dog if a strong rank order develops because these dogs are not genetic relatives. Such family dog packs may also function peacefully on the basis of friendly relationships, if they have been socialized to other dogs at a young age and have enough time under the owner's surveillance to get to know one another.

Above *Only packs of animals are successful in hunting for big game. After the killing, each member of the pack gets its share, although some order applies.*

Left *Hunting is often a collaborative effort for wolves. It starts with the pack mates searching the area and looking for any signs of prey.*

Hierarchy & Cooperation

It has long been argued that our companion dogs tend to develop a strict rank order and continuously strive for dominance in their "pack"—similar to the social hierarchy of wolves. Recently, however, plenty of evidence has accumulated showing that these arguments are flawed, mostly because the observations have been of captive wolves, providing a false picture of social organization in wolves.

Although some form of hierarchy is important and adaptive in most social species, its nature is heavily influenced by the ecological and social constraints present in the specific environment.

Generally, a very strict hierarchy also inhibits the emergence of cooperation.

ZOO WOLVES & WOLVES IN NATURE

The composition of captive wolf groups is very different from that of a natural pack because wolves in zoos and wolf-parks are not close relatives and there are too many adult individuals sharing a limited space; that is why violent rivalries for leadership have been observed.

In contrast, a natural wolf pack is more likely to be an extended family including the breeding pair and some

Above right Following a specific protocol when we feed our dogs is important not because of some "dominance"-related reason, but simply because it is a proper opportunity to teach them self-control.

Below The classical notion of the strict hierarchical structure of wolf packs has been based on observations of captive wolves.

generations of their pups, where the offspring disperse before they reach sexual maturity. The wolf's natural social behavior is now known to be based primarily not on aggressive interactions but on affiliative family bonds. Being a family also ensures that wolves display high levels of cooperative behavior, especially when they venture out to hunt or organize the feeding of the pups.

THE DOG'S POSITION IN DOMESTIC SETTINGS

Through the process of domestication, the importance of direct aggression in the maintenance of rank order between humans and dogs has gradually diminished, with dogs becoming more and more dependent on humans, thus ensuring their subordinate position. The development of individualized attachment relationships with humans and decreased levels of aggression formed the basis for cooperative behavior.

Research shows that companion dogs living in human families generally do not strive to fight their way to the "alpha" position; on the contrary, they are keen to cooperate with their owners even in complex tasks where the interaction

A GOOD LEADER

The outdated and false theory of the "alpha male" suggested the owner should take on the role of pack leader by communicating steady and unambiguous signals of his/her dominant rank. However, due to decades of misuse, the "alpha" terminology nowadays falsely implies a rigid, force-based dominance hierarchy.

When interacting with our companion dogs, we do better to try to model a family pack, where contests are rare and the parents are natural leaders of the younger ones. Most dogs need an experienced, self-confident leader, who manages the group by taking the initiative, rather than a rude boss.

cannot be based on a dominance relationship (in the case of service dogs, for example).

The notion that the owner can only be leader or follower is incompatible with cooperation; by its very nature, cooperation is based on an effective bond and a trusting social environment. This does not mean that wolves or dogs do not display social dominance, but it is highly situational, and in the case of dogs it varies greatly between individuals.

Above *The wolf pack is a typical family. The effect of positive interactions on the social life of wolf packs has long been neglected.*

Affiliation & Aggression

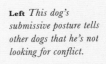

Social relationships between dogs and between humans and dogs are influenced both by agonistic encounters and affiliative interactions. Friendly interactions, active submission, play, reconciliation, and consolation all help avoid conflict and contribute significantly to the stability of the group.

Though most aspects of the social behavior of dogs can be understood by observing them interacting with

conspecifics, domestication has changed the dog's species recognition system. The human–dog relationship is based on dogs' spontaneously emerging attachment, and in the vast majority of cases companion dogs show exclusively affiliative behaviors during our interactions. With this in mind, the use of old concepts for the owner–dog bond seems imperfect and the idea of considering our relationship as a friendship—though not necessarily a symmetrical one—has recently been put forward.

AGGRESSION AS COMMUNICATION

The main function of aggression is to divide limited resources among group members with minimal conflict. Aggressive behavior in social species consists mainly of displays used for communication. Consequently, aggressive behavior does not necessarily mean physical aggression. Dogs indulge in a fight only if there is no better (that is, less costly/dangerous) solution. Conflicts within the group are solved by the use of agonistic signals to avoid the escalation of the situation.

Left *This dog's submissive posture tells other dogs that he's not looking for conflict.*

Above *Aggressive behaviors are often fear-related responses, thus neither counter-aggression nor reassurance are appropriate reactions. Attracting their attention to something (such as food or a toy) and praising them when they stop the aggressive displays is the best strategy.*

HUMAN-DIRECTED AGGRESSION

During domestication, the tendency of direct aggression toward humans has gradually diminished, and interactions have been increasingly influenced by affiliative and cooperative tendencies. The general lower level of human-directed aggression emerged in parallel with increased controllability of competitive behaviors. Nonetheless, dogs' human-related aggression poses serious public health and animal welfare concerns even nowadays.

DOMINANCE AGGRESSION

In a hierarchical social structure, fighting to secure a higher rank is the typical case for dominance aggression. Dominance aggression in dogs has been linked to nonauthoritarian owners, lack of obedience training, spoiling or anthropomorphizing the dog, and not using physical punishment.

Recently, this view has become strongly debated because, in general, hierarchical relationships both among companion dogs (belonging to the same group, that is, in daycare facilities) and between dogs and humans are more symmetrical, closer to an egalitarian connection. However, any social species that has the ability to form hierarchies also has the potential to strive for higher rank. The dog is no exception to this rule. However, the occurrence of dominance aggression in companion dogs can have many causes and is considered as problem behavior in most cases.

FUNCTION OF AGGRESSION

The function of dog aggression (or aggressive communication) is to obtain or protect some resource such as food, a toy, or a sleeping place, while the target of aggressive behavior can be either dogs or humans. In social species, aggression differs depending on whether it occurs within a group or between groups. Dog-related aggression occurs in two main contexts—between dogs in the household, and directed at non-household dogs. Females tend to initiate more household aggression, whereas males are more prone to attack non-household dogs. Same-sex pairs, especially females, are more frequently involved in household aggression than opposite-sex pairs.

RESOURCE HOLDING POTENTIAL

Exhibiting resource guarding is a perfectly normal and appropriate behavior among conspecifics. However, when dogs also defend their "ownership" aggressively toward humans, we need to interfere to avoid injuries, even if this behavior is still considered acceptable. Modifying the behavior of a heavy resource guarder is possible but can take a long time and experience, and some individuals will never be perfectly trustworthy.

VARIABILITY OF AGGRESSIVE TENDENCIES IN DOG BREEDS

Dog breeding for work purposes has markedly changed aggressive behavior in dogs. In some cases, such tendencies were selected for, while in others there was a selection against it. However, the lack of selection of some of these traits in modern dog breeding has increased the frequency of aggressive behavior when it should have been controlled for. There is relatively widespread aggressive behavior in some lines of golden retrievers, a breeding issue exacerbated by the assumption that all golden retrievers are friendly.

Some breeds, such as the English bulldog or mastiff, are often regarded as "dangerous" only because of the original utilization of their ancestors as fighting dogs. However, the modern breed shares very little in terms of behavior with the old bull-baiting type.

Differences in breed-specific aggressive behavior in different countries are probably due to different breeding legislations—for example, the breeding legislation for German shepherd dogs differs between the UK and Germany. Breeds found to be the most aggressive in some countries may be friendlier in others.

Left *Even the friendliest and calmest dogs may tend to guard their possessions. This is part of their natural behavior but can be successfully resolved.*

According to surveys, dog aggression within the household is less frequent among toy and sporting breeds, and more frequent among herding breeds. Cases of non-household aggression are more prevalent among terrier breeds.

Among English springer spaniels, conformation-bred dogs have been found to be more aggressive toward both humans and dogs than field-bred spaniels. However, the opposite pattern has been observed for owner-directed aggression among Labrador retrievers. This indicates that higher levels of aggression are not attributable to breeding for show per se.

Socialization and training can definitely modify genetic predisposition, but do not wipe out significant differences. It is impossible to create a military dog from an English setter, while there are many individuals from specific breeds that do not make good pets. Importantly, however, breed-related differences in aggression do not provide any reason for any breed to be discriminated against.

Toy

Sporting

Herding

More frequent aggresion

Herding

Less frequent aggresion

English setter

HAPPY TAIL WAGGING— ANGRY BARK?

Dogs wag their tails and bark for many reasons, and it would be an erroneous oversimplification to claim that they convey one unambiguous message. Both types of signal reflect some sort of excitement, and we should not interpret them in isolation from the other components of body language.

While dogs wag their tails mostly in greeting situations, and during friendly encounters, tail wagging can also serve as a submissive signal in conflict situations. The common misconception of happy tail wagging could even lead to dangerous situations, because a high, stiff wag can communicate tension or hostility. The messages from dog barks can also be variable, from threatening to conveying despair, fear, or even happiness.

Above *Aggressive tendencies within the household can vary from the level of aggression shown toward strangers.*

Below *A tail held high should not be interpreted as a signal of dominance toward an unfamiliar dog because they have not yet established any rank order.*

Territorial Behavior ～🖊

Dogs living in human society have little need to compete for resources and they do not have to protect territories. In spite of this, elements of territorial behavior are still present in dogs even though they have been affected by domestication and vary a lot with breed.

TERRITORIAL AGGRESSION

There is scarce information about the territorial behavior of family dogs, especially as their movement is limited and controlled by their owners. Although some dogs do seem to defend their living environment (house, yard), this behavior was favored only in the selection of some working dog breeds (such as livestock guard dogs). Instead of long-range howling, barking became the main acoustic signal of dogs in territorial defense. Barking may also have a recruiting effect, encouraging other group members to join the signaler. As such, dogs bark not just at an intruder into their living area, but at any approaching individual. Nearby dogs may join in to form a barking chorus.

SCENT MARKING

It is not known whether dogs rely on odorous cues from the feces as a form of olfactory communication. However, the presence of an anal sac and the composition of the excretion of the anal glands are individual specific and suggest a communicative function.

In contrast, dogs clearly pay attention to urine marks. Males (and less frequently females, too) display a raised leg during urination for marking elevated places. This raised-leg posture is also used sometimes in the absence of urination, in the presence of other dogs, suggesting that it is a dominance or territory-ownership signal.

Over-marking is also a common behavior in both sexes, indicating that this behavior is the remnant of territorial defense, which is also observed

Right Dogs use different postures to deposit urine scent marks but raised-leg urination is the most common form in males. This posture helps to elevate the mark to nose height.

Below right The ideal livestock guardian dog is extremely territorial and aggressive to strangers but mild and docile toward livestock. It can move calmly among the sheep without disturbing them.

Wolf packs keep large territories, the size of which may change dynamically with the season and the availability of prey. Size also depends on the habitat: In woodlands a relatively small area (below 380 miles^2/1,000 km^2) can provide the resources for one pack, while in northern tundra regions, some territories expand to more than 1,160 miles^2/ 3,000 km^2. Wolves protect their territory by scent marking and howling. Some scent marks (urine and feces) are left within the territory but occur more frequently along the borders at approximately every 275 yd. (250 m), and the marks are effective for 2–3 weeks in repelling members of the neighboring packs.

more frequently in males. Males tend to scratch the ground after elimination with their hind legs, kicking back the soil, possibly to spread the scent more effectively. Scent marks, besides carrying individual information, may also inform males about the receptivity state of female dogs in the area.

VARIATIONS AMONG BREEDS

Individuals belonging to dog breeds selected for livestock guarding and herding display enhanced sensitivity toward intruders in their living area, and may respond with elevated aggression toward other dogs, humans, and some other animals, including wolves. In contrast, in the case of breeds that were selected for purposes such as hunting, the trait of tolerance toward human strangers and conspecifics was probably a more favorable trait.

Courtship & Mating ~

There is very little knowledge of the courtship behavior of wolves in nature, and observations of captive wolves do not provide a true picture. Wolves live in monogamous pairs and their bond lasts over many years. This is rather different from the situation in dogs, including free-ranging dogs, where monogamy is relatively rare and both males and females mate with several partners during the mating season.

COURTSHIP BEHAVIOR IN CANINES

It is likely that all canine species, including companion dogs, share a common pattern of courtship. Courtship behavior begins with the male dog starting a kind of dancing action around the female, intermittently lowering his front part and wagging his tail. He may nip at different parts of the female's body, preferentially at her face, neck, and ears. Attempts at mounting are first shown toward the side of the female, and if she shows willingness to mate then the male mounts her from the back. If the female finds the male's courtship attractive, she may take up a submissive posture accompanied by whimpering sounds, and pull her tail to the side. She exposes the

Courtship

Mounting attempt

Copulation

Chosen mate

Copulatory tie

Above *The steps of mating are similar for all canids. The most peculiar behavior is the copulatory tie, when the male and female stay together for quite a few minutes.*

Left *All members of the* Canis *genus are strictly monogamous, which means that typically only the oldest breeding female sires offspring. This does not exclude courtship or even mounting among different individuals in the pack.*

Below left *Stray dogs are characterized by a variety of reproductive systems because unrelated individuals live in different types of groups. Females may give birth to puppies of different fathers.*

genitals to the mounting male and stands firm to hold his weight.

Only a small proportion of the courtship action ends in copulation, and the courtship ceremony may be shortened significantly if there is a strong bond between the individuals, as is the case in wild canines living in families.

COPULATORY TIE

The small bone in the penis facilities its insertion into the vagina where the erection occurs. After this, the erectile tissue of the penis becomes infiltrated with blood. This way the penis gets locked in the vagina, and despite the fact that the ejection occurs relatively soon after the insertion, the two animals stay joined to each other for the following 5–20 minutes. This copulatory tie is specific for canine males, and its function is not known, although it may ensure that the semen of the male gets on its course toward the ovaries in the absence of any interference from other males.

REPRODUCTION IN FREE-RANGING DOGS

Observations show that free-ranging dogs engage in various forms of mating. Although monogamy, similar to that observed in wolves, may occur, it is more common that one female mates with many males (polyandry) or both females and males mate multiple times (promiscuity). Occasionally, males not belonging to that group have been observed to mate forcefully with females (rape). The diversity of female–male contacts also suggests that courtship behavior in these dogs is also distorted. Free-ranging female dogs are quite selective—in 90 percent of cases they copulate with preferred males.

REPRODUCTION BEHAVIOR IN PUREBRED DOGS

In most purebred dogs, reproduction is arranged by the breeders. Based on observations in wolves and free-ranging dogs, it is highly likely that this method negatively affects the reproductive behavior of dogs living in human societies, and this situation is exacerbated by the use of artificial insemination. Males are not selected for their ability to court females, to be able to deter other males, and to perform a successful mounting. Females are not allowed to choose males based on their preference, which, in natural situations, is often based on genetic compatibility.

Development ～ℓ

Much of our knowledge about the development of dog behavior dates back to the 1950s and 1960s, when John Scott and John Fuller carried out massive studies over 13 years at the Jackson Laboratory in Bar Harbor, Maine. The aim of these studies was to address the genetic influence on the development of social behavior.

DEVELOPMENT IS A HETEROGENEOUS PROCESS

Dog puppies are born blind and deaf; they are not able to walk, can barely crawl, and do not survive without their mother's care. In the subsequent weeks and months, they grow rapidly in size and develop the abilities and skills they need as adults. The size of newborn puppies differs depending on the size of the breed, so the duration of the physical development of dog puppies varies greatly, depending on the size the dog reaches as an adult. For very small dogs it may take approximately 6 months to reach their adult size, while for giant breeds it may take 18 months. There are also differences in the timing of development between breeds, with some skills and behaviors emerging much sooner in some breeds than in others.

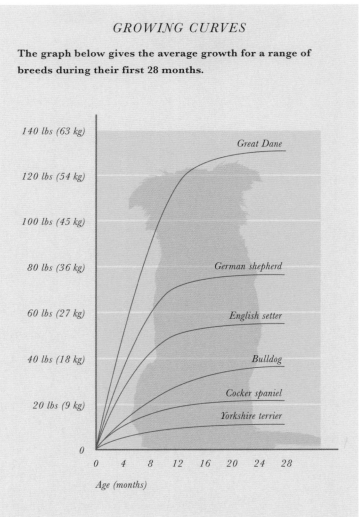

GROWING CURVES

The graph below gives the average growth for a range of breeds during their first 28 months.

DEPRIVATION FROM SOCIAL PARTNERS

It is well known that early experiences can greatly affect the later behavior of dogs. In some early experiments researchers deprived dog puppies at various ages of human contact. Dogs that had never experienced humans during their early development showed marked avoidance toward them, and this behavior could not be alleviated by subsequent socialization. This explains why many feral dogs that do not spend time with humans as puppies keep avoiding people later in life. However, dogs are special because even a very little social exposure, up to a few hours per day, may develop their preference for humans.

DEVELOPMENTAL EXPERIENCE

From birth to death canines undergo a series of changes in their physical, ecological, and social environment. For example, a few weeks after birth, from the safety of the small and confined space of the litter, puppies are gradually exposed to richer and more stimulating surroundings. Puppies learn to recognize individuals, to form affiliative relationships with some, and to avoid others. Dogs' social environment is particularly rich and complex because it includes not only conspecifics but also members of another species: humans.

Opposite top *A Great Dane puppy sits between its mother's legs. For giant breeds it may take 18 months to reach their full adult size.*

SENSITIVE PERIODS

During sensitive periods the puppy is exceptionally quick to learn about particular stimuli in its environment. The experience gained during this period is thought to have a great impact on future behavior. If the dog misses specific inputs, it may develop behavior malformations. Lack of experience with other dogs may lead to inappropriate behavior, including fear or aggression when encountering a conspecific.

Puppies & Early Life

Puppy development has traditionally been divided into stages or periods. Although there is some conceptual advantage in this division, the development of behavior is not a precisely defined step-by-step process, but rather a continuum of changes throughout the dog's life. Start and end dates should be considered as flexible, especially as dog breeds vary greatly with regard to the time when they pass through a specific developmental stage. Changes in the behavior of puppies are caused by a complex interplay between genetic and environmental factors so their timing will vary.

The timing of the developmental stages in wolves and dogs are similar. However, wolf pups and dog puppies gain very different experiences as they grow up. Thus, the differences between wolves and dogs are not only genetic in origin but are also strongly influenced by environmental input.

NEONATAL PERIOD (FROM BIRTH TO DAYS 10–12)

Wolves usually dig an underground den and give birth there. The pup's ability to perceive the world at this age is restricted to tactile and olfactory stimuli. The mother and littermates are the only sources of physical interaction, including stimulation of olfactory receptors in the nose and tactile receptors around the mouth, important for the suckling activity.

Dog puppies experience a different situation: Humans usually create artificial "dens" for the puppies that are better lit and contain more stimuli than those of wolves. Already at this age puppies can learn about odors and tactile stimuli. Olfaction-based learning might be particularly robust. Although the puppies' locomotor abilities are very limited, they wrestle with their littermates for position at the nipples.

Above *In the neonatal period puppies spend most of their time sleeping close to one another. At this age they rely mostly on the olfactory senses.*

Below *Yorkshire terrier puppies at different stages of development; they grow relatively slowly despite their small size. Their cuteness should not deceive owners – they need to be socialized to humans as early as possible.*

Birth to 10 days   *14 days, eyes open*   *3-4 weeks*

WHY MALE DOGS ARE NOT GOOD FATHERS

Wolves are monogamous, so the male wolf feeds the female by taking food to her at the den. Later, they also provide cubs with food from the hunt and regurgitate it for them. Human care has interfered with this behavior in dogs because the male dogs are not present when puppies are born, and are not involved in the parenting. This may explain why many male dogs show a negative reaction to puppies if they are not socialized appropriately to them.

Right *In wolves, both the female and the male, as well as older sisters and brothers of the pack, provide care to the puppies and juveniles.*

6 weeks   8 weeks   10 weeks   12 weeks

TRANSITION PERIOD
(FROM DAY 13 TO DAYS 20–22)

This period is characterized by rapid development of perceptual abilities. It starts with the opening of the eyes and terminates when puppies start to hear. As there is a big variation in the timing of those events in different breeds and these events are independent of each other, there is also a big variation in the duration of this period.

During the transition period the coordination of movements improves rapidly, allowing the puppies to perform more complex behaviors when interacting with others. The puppies start to jump, play using biting, and wag their tails in social interactions. At the same time the stimulations between mother and puppies start to decrease.

SOCIALIZATION PERIOD
(FROM WEEK 3 TO WEEK 12)

Wolf pups emerge from the den at the beginning of this period. At this time, they are exposed to a richer social environment and meet all members of the pack. Through interaction with them, puppies improve their motor and social skills. During this period puppies are weaned and learn to beg for food from the others, for example by licking the corner of the adult's mouth, which elicits vomiting in the adult wolf. Adult wolves may also carry uneaten pieces of meat and this provides the opportunity to experience food sharing and competition for food. In turn, this situation promotes learning about hierarchy.

For family dog puppies, the human family provides the most influential social experience. While they are in the litter, dog puppies also form transient hierarchical relationships with their sisters and brothers. This period is the most important for learning about social relationships and in

Above left *The transition period starts when the eyes open, and puppies undergo a rapid development of their perceptual abilities.*

Above *During the socialization period puppies improve their motor and social skills through interacting with peers. This is the most important period for puppies to learn about social relationships.*

experiencing how to integrate into a social network. There is also evidence that already at this age puppies learn from both dogs and humans by observation.

Puppies are usually adopted around their eighth week, and lose close contact with littermates and other dogs. Many of them do not have the chance to learn how to become a member of a social network of conspecifics. Owners can give their puppy appropriate experience with conspecifics by taking it regularly to puppy classes or allowing it to interact with other dogs in different situations.

JUVENILE PERIOD (FROM WEEK 12 TO SEXUAL MATURITY)

At this stage wolf pups are beginning to participate in the social life of the pack. To start with, they wait around the den or other places for the adult pack members to return from hunting. If the hunt was successful, the returning wolves regurgitate some food to the juveniles. If pups gain enough strength and endurance, they also join the hunt. During this period they learn a lot about how to participate in a hunting team and how to navigate in their territory, and they assume a specific role in the family. In nature, two-year-old wolves are considered as adults; most of them leave the pack by this age to form a new family in a new territory.

Many dogs spend this period without conspecifics in human families. Thus, it is very important to provide them with frequent contact with other dogs through puppy classes or in other ways. By 3–4 months of age puppies start to develop attachment to specific individuals, which

mainly manifests as staying in proximity to this figure or running to it in stressful situations.

This period extends until dogs reach sexual maturity, which usually occurs between 9 and 18 months of age, with a large variability between breeds. In some breeds, females may have their first estrus as early as 5–6 months. Although dogs mature sexually earlier than wolves, they do not display a fully adult behavioral repertoire until later. Some breeds develop fully adult behavior only at around 2 years old.

Above *The juvenile period is important for getting a lot of experience with unfamiliar humans and dogs. Juvenile dogs living in apartments often get limited contact with conspecifics. Owners should provide frequent opportunities for them to interact with other dogs.*

BREEDS MAY DEVELOP AT DIFFERENT SPEEDS

The timing of development of skills and the emergence of behaviors varies between breeds. Cocker spaniels open their eyes around day 14, while most fox terriers do so only a few days later. In contrast, fox terriers start to hear sooner than cocker spaniels. Similar differences also emerge in puppies' behavior. Huskies start walking much earlier than German shepherd dogs or Labradors.

Socialization

From the time when a puppy is adopted, usually at around 8 weeks old, it is no longer with its siblings and mother and its experience of being a member of a social group of conspecifics is rather limited. Dogs living with humans need to learn to behave agreeably with unfamiliar individuals of their own species and also with unfamiliar humans. The duration of the period in which dog puppies can develop social relationships is relatively long compared to wolves.

LEARNING TO BE SOCIABLE

The period between week 3 and week 12 is very important for developing the ability to form social relationships with others. Breeders and owners have the task of exposing the puppy to pleasant social stimulation, allowing puppies to interact with a variety of people of different ages, including children, women, men, and, when possible, people of different ethnicities and abilities. They also need

Opposite top *Puppies typically show submissive behavior with adults. But a well-socialized adult dog will respect the puppy's "special" social status.*

Below *A puppy interacts with other dogs during a puppy class. In the socialization period it is fundamental for puppies to learn about social interactions.*

to learn about other objects that will be part of their environment, such as elevators, cars, cell phones, and vacuum cleaners. Puppies that do not have such experiences in this important period may react with fear to specific categories of people. For example, puppies who are only exposed to women may later show fear of men.

IS IT A DOG?

Similarly, puppies need to socialize with other dogs of different ages and of different physical appearance. Dogs of breeds with extreme features, such as a short snout or long floppy ears, might even hardly be recognized as dogs by a puppy without proper previous experience. Taking the puppy to puppy classes where other dogs of various breeds can be encountered in a controlled and pleasant environment can provide the right social experience.

AT WHAT AGE SHOULD A PUPPY BE ADOPTED?

Bearing in mind the importance of the role of the mother and the siblings in developing a correct behavior repertoire with conspecifics, there is no rush to separate puppies from their native family. It is generally recommended that a puppy is allowed to stay with its mother and littermates until at least 8 weeks. Choosing a puppy from a breeder who pays special attention to the socialization of the puppies, by exposing them to various social stimuli during the early weeks, is also recommended, especially if the puppy is adopted at an older age.

EARLY ACCLIMATIZATION

Being carried in a car is a typical example of something most dogs can get used to, if they are properly exposed to it during the socialization period. If owners take the puppy for short trips every day, the puppy will become accustomed to traveling in the car. However, if this is not done regularly during this period, dogs may later need to undergo specific procedures in order to learn to cope with the strange situation of being confined in a small, moving, and noisy space.

Play Behavior in Dogs ～℮

In social species, play represents one of the most complex interactions between two members of a group. The behavior elements displayed during play are actions borrowed from various other behavioral contexts (including agonistic and predatory), but they are modified versions of the original actions and can be combined in novel ways. At the higher level of complexity, during play the partners need to cooperate and adjust their actions in order to achieve their common goal of playing together.

THE FUNCTION OF PLAY

Social play appears to serve many functions in juveniles, improving physical fitness and motor skills, and helping the puppy to learn social skills such as bite inhibition. Social play allows dogs to practice and combine actions they will use later in their lives, and in general play also prepares them for the unexpected. In dogs (and wolves) social play is not limited to juvenile individuals—adults play too— so one main role of play is to maintain the social cohesion of the group.

One of the biggest questions in the study of dyadic play is whether intentional descriptions are appropriate when we interpret the dog's behavior. Play can take many forms, from simple learned play, such as ball fetching, to pretend play, where the dog displays signals indicating an inner state (aggression, for example) that is not real, such as when the dog pretends he is defending an object.

TUG-OF-WAR & OTHER COMPETITIVE PLAYS

There has been much debate about playing tug-of-war with our dogs. Some argue that such situations trigger aggression or assertive behavior, warning people that at least they should not let their dogs win in the tug-of-war game. There is no evidence to support such opinions; in fact, the situation is just the reverse.

DOG PLAY SIGNALS

Play signals have several functions: They clearly distinguish such interactions from real competitive situations, and they serve to initiate play and also to synchronize the actions of the partners. Play signals can be highly variable and can include open mouth display, high-pitched barking, bounding over to the other dog in an exaggerated manner, a bowed head, pawing, or exaggerated retreat. Barking used as a play signal is specific to dogs; it is absent in the play of other canines.

The best-known, highly stereotyped play signal in dogs is the play-bow. It not only conveys the playful intent but it is also used after ambiguous behaviors (such as a playful bite or snap) to display the dog's willingness to continue the interaction. When the play partners are familiar with each other, bows most often occur after a brief pause with the aim of reinitiating play.

HUMAN PLAY SIGNALS

Humans are also quite successful in initiating play in dogs by using different signals. These can be borrowed from human communication, or copied from the signals displayed by dogs. A person can successfully evoke play by presenting a bow or a lunge, especially if he or she also uses high-pitched vocalization. Because anything that happens during play is "not real," contrary to common belief, dogs do not apply their play experiences to the dominant/submissive nature of social relationships. Running together for a while, then chasing and running away in turns can be the best game with most dogs.

Above *Play signals not only convey the dog's playful intent, but also promise that what comes next is not meant to be serious. The so-called "play bow" is typical for canids and has been shaped by evolution.*

Left *For active dogs, chasing a ball or jumping for a flying disc can substitute the experience of joint running or wrestling with the human play-partner. Such activities represent ritualized forms of group hunting.*

DIFFERENCE BETWEEN REAL & PRETEND

Dogs commonly play chase and rough play, which closely resemble predatory and fighting behaviors: They tackle, bite another dog's neck, hip check, bare their teeth, mount, rear up, stand over and do chin-overs, bark, and growl.

Play fighting is significantly distinct from real fighting, and not just because it may start with play signals. When playing, dogs inhibit the force of their bites and voluntarily give their partner some advantage by, for example, running slowly and letting themselves be caught or lying down and rolling on their backs. These self-handicapping behaviors would never happen during real fights but help initiate and maintain social play between partners of different sizes or strengths. Dogs often apply the same behaviors when playing with young children.

THE PARTNER MATTERS

No single conceptual framework can be applied to both dog–dog and dog–human play because they are motivationally distinct. The opportunity to engage in dog–dog play in multi-dog households does not decrease the dogs' motivation to play with their owners too, which would be likely if the two types of play were motivationally interchangeable.

Dogs are more likely to give up on a competition, or to show and present a toy to their human play partner than to another dog. They are also more interactive and less likely to possess the toy when playing with a person.

The dog–dog and dog–human play seem structurally different, which suggests

there is no reason to assume that the consequences of dog–dog play can be extrapolated to play with humans.

OBJECT PLAY

Animals rarely use objects as means of play, apart from between dogs and humans, when this is the most common form of play. In play situations with a human the behavior of well-socialized family dogs is influenced more by their motivation to play than by the familiarity of the play partner or their possible general tendencies for cooperative or competitive behaviors.

COMPETITIVE PLAY

It was once assumed that competitive games increase agonistic tendencies in behavior, suggesting an effect of play activity on later sociability with partners. It turned out, however, that competitive games do not increase aggressive tendencies in real-life situations. On the contrary, it seems that the type of game dogs prefer to play depends on whether they have a cooperative or competitive personality.

Over time dog and owners develop a routine of games, and dogs do not generalize these behavior routines to other, functionally different situations. Thus, it is very important, starting during puppy age, that the dog gets many opportunities to play with other dogs and also humans. Play is one of the best ways to improve the physical and social skills of dogs, and it also facilitates people's understanding of their companion. A day without play is a lost day!

Below *Rough and tumble is a form of play-fighting for dogs and only rarely escalates to real conflict. To ensure harmless play, dogs need to get the necessary experience as puppies.*

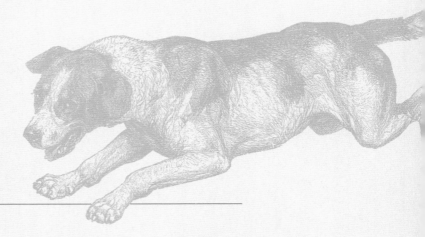

The Old Dog

It is not just human life expectancy that has increased in recent decades. The protective human environment has increased the life expectancy of dogs too. However, very little attention has been paid to old dogs and not much is known about the actual prevalence and risk factors of age-related changes in dogs.

THE DIVERSITY OF LIFE SPAN IN DOGS

The maximum life span recorded for dogs is around 22–24 years. Purebred dogs' mean life span ranges from 5.5 to 14.5 years, whereas for mixed-breed dogs it is about 13 years. However, only 15 percent of dogs die because of old age. The most frequent cause of mortality varies from breed to breed, but dogs die mainly from cancer and cardiac problems.

AT WHAT AGE SHOULD A DOG BE CONSIDERED OLD?

It is difficult to establish a general threshold for aging in dogs because of the huge variability in life spans of breeds. In the beagle, for example, it is common to differentiate five life periods: young adult (1–3 years), adult (3–6), middle-aged (6–8), old (8–10), and senior (11-plus).

However, the majority of dog breeds have a shorter life expectancy than the beagle (see box opposite), so these values should be adjusted to each breed based on their expected life span.

MAIN FACTORS INFLUENCING THE LIFE SPAN

The life span of dogs mainly depends on body size. The larger dogs (154–176 lb / 70–80 kg) live up to an average of 7–8 years, 6 years less than dogs weighing 22–44 lb (10–20 kg) do. Early rapid growth is probably the major factor behind faster aging in these large dog breeds. The role of inbreeding among breeds must also be taken into account because, among dogs of the same body size, mixed breeds generally live longer. Interestingly, dog breeds with a more trainable character tend to have a longer life span. These dogs probably take fewer risks but are also less stressed, and therefore healthier, because of their docile manner.

Below *Beagles older than 11 years are considered as seniors. However, individuals of larger breeds live only up to 7–8 years.*

COGNITIVE DECLINE

Similarly to elderly humans, dogs also show age-related cognitive decline as time goes by. About 30 percent of 11- to 12-year-old dogs and 70 percent of 15- to 16-year-old dogs display cognitive disturbances that correspond to human senile dementia. These dogs exhibit spatial disorientation, social behavior disorders (for example, they are not able to recognize family members), repetitive (stereotypic) behavior, apathy, increased irritability, sleep–wake cycle disruption, incontinence, and reduced ability to accomplish tasks.

Extrinsic factors may protect against rapid cognitive decline. In a study of aged laboratory beagles, groups undergoing an enrichment program (including increased exercise, environmental enrichment, and cognitive enrichment) performed better in learning tasks than the control group. This improvement was greater than that following the application of an antioxidant fortified food.

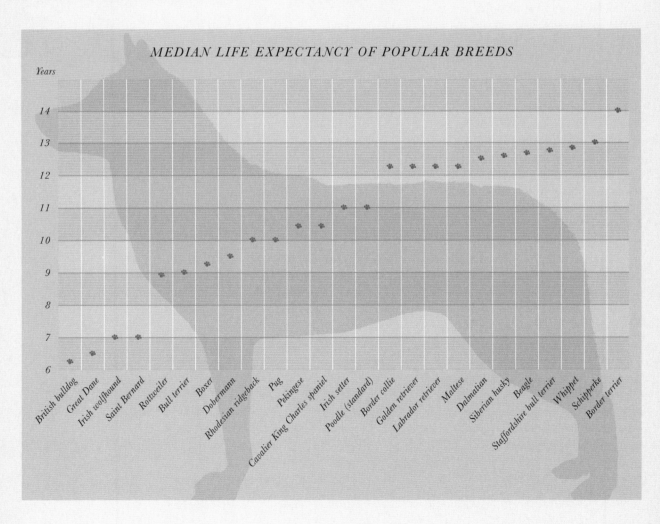

MEDIAN LIFE EXPECTANCY OF POPULAR BREEDS

Communication Between Dogs & Humans ~𝒬

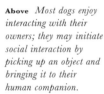

It is beyond question that animal communication is a complex phenomenon that often gives the impression that the observed behavior is brought about by high-level cognitive mechanisms. Therefore, it is no wonder that, when thinking of animal communication in general and dogs' communication skills in particular, one can fall into the trap of anthropomorphism and endow animals with human-like mental abilities.

Admittedly, it is not easy to talk about dog–human communicative interactions without using anthropocentric phrases like "providing information," "being instructed," or "wanting to please." Importantly, however, communication in nonhuman species is in most (though not all) cases cognitively less complex than it might seem at first glance.

Signal production by senders and receivers' responses are typically based on flexible mental representations (reflecting the actual inner state) rather than manifestations of complex insightful cognitive processing. If, for example, a dog growls and snaps when another dog approaches his food, the adaptive function of this possessive aggression is to gain control over a resource. This is the dog's way of saying, "I am angry! Clear off!" This evokes relevant responses from the competitor without gaining insights into what is going on in the other dog's mind when he protects his food.

THE EVOLUTION OF COMMUNICATION: BASIC CONCEPTS

Communication is an interactive process during which a signaler displays and a receiver responds to a signal. Signals are perceivable behaviors (or bodily features) that have the potential to change the behavior of a receiver in a way that is beneficial to the signaler, not excluding benefits on the part of the receiver.

Communicative signals passing through various sensory processes (visual, auditory, olfactory, and tactile) may evolve from preexisting behaviors that already have some value to the potential receivers. If the receiver's response evoked by such informative behavior is beneficial to the signaler, then, on the evolutionary time scale, the behavior becomes gradually transformed into a communicative signal by increasing conspicuousness, stereotypy, and separation from its original function.

This process is called evolutionary ritualization, during which the behavior evolves to a signal that elicits the most appropriate response from the receiver.

Above *Most dogs enjoy interacting with their owners; they may initiate social interaction by picking up an object and bringing it to their human companion.*

Opposite top *Dogs readily attend to human-given cues, and they show a particular preference for face-to-face interactions and eye contact with humans. But their attention toward humans depends on their socialization and relationship.*

Opposite below *Slapping his front paws down onto the ground repeatedly is the dog's typical way of signaling that he wants to play.*

Although the original function of hair bristling is to regulate body temperature, hair bristles on the back and shoulders also make dogs appear stronger and bigger than they really are. Virtual body size is an important informing cue in conflicts, and thus hair bristling has become ritualized as a communicative signal indicating an aggressive behavioral state that is produced in a wide variety of contexts.

Ritualization may also take place at a developmental timescale. This latter process is called ontogenetic ritualization, during which individuals mutually shape their behaviors over repeated instances of social interactions and the signaling function of certain behaviors is shaped through individual learning.

WHAT DO ANIMAL SIGNALS MEAN?

Communicative signals can be seen either as indicators of motivational processes or they may reflect external events in the environment. These signals are independent from the sender's internal state, and carry referential meaning (as human words do). Research suggests that animal communication is typically motivational in nature, which means that signals reflect the internal state of the sender, so aggressive signals, for example, may be linked to particular changes in the physiological state. Importantly, however, receivers may make inferences about external events from perceiving the signal.

Here is an example. Growling at an approaching dog in a food-guarding situation reflects the motivational state of the sender. However, dogs growl in many different situations (even during play) and

HUMAN LANGUAGE & ANIMAL COMMUNICATION: KEY DIFFERENCES

Human linguistic communication is symbolic in nature, and the messages are coded in digital form (in terms of the presence/absence of a linguistic signal) rather than in analog form (in terms of gradual change in the signal). That is to say, each word has a specific set of meanings and one word does not graduate into another. Moreover, humans generate an infinite number of possible messages with a finite number of signals (phonemes).

In contrast, nonhuman communicative signals can be weighed on an analog scale. Animals send a set of potentially continuously varying signals, such as growls of varying intensity, and they convey information about a limited set of events or states. Animals display signals only as a response to inner or environmental changes—danger needs to be present for an individual to signal danger, for example—while humans may signal spontaneously.

the acoustic structure of the growls is context-specific. Thus, receiver dogs that know something about the context-specific features of growls could infer, based only on the auditory cue, which situation the signaler dog has encountered even though its growl is purely motivational.

THE EVOLUTIONARY ROOTS OF DOG–HUMAN COMMUNICATION

Humans utilize a set of uniquely complex communication skills that are very different from those of canines. Thus, to be successful in the human community, during the process of domestication dogs had to be selected for an enhanced ability to fit into the human communication network. Research suggests that domestication has led to:

1. Profound expansion of the stimulus range that dogs are able to attend to;
2. The emergence of less rigid and more diverse communication behavior patterns;
3. A particular preference for pivotal aspects of human communicative exchanges such as face-to-face interactions and eye contact with humans.

Right *Most dogs readily understand the pointing gesture, and some dogs are especially good at picking up the direction indicated.*

Below *The nose-to-nose approach is a typical canid way of interacting. Many humans are keen on such contact but it may lead to injuries, especially with an unfamiliar dog.*

Adopting a low posture

Tail tucked under

Averted gaze

THE PRINCIPLE OF ANTITHESIS

Signals with opposite meanings are often conveyed using opposite visual and acoustic signals. This tendency is present in both animal signaling and human nonverbal communication. For instance, assertive dogs display a high posture (standing tall, tail up) while a withdrawn dog (left) tends to adopt a low posture (standing low, tucked tail).

Thus, dogs have acquired the ability to adjust their communication to the human environment. In particular, changes in the species-specific communication system in the dog contribute to their ability to perceive and respond appropriately to human signaling (for example, pointing and other referential gestures) and produce functionally human-like communicative signals (such as gaze shifts between people and objects).

THE EMERGENCE OF INTERSPECIES COMMUNICATION

There are five basic conditions to enable interspecies communication.

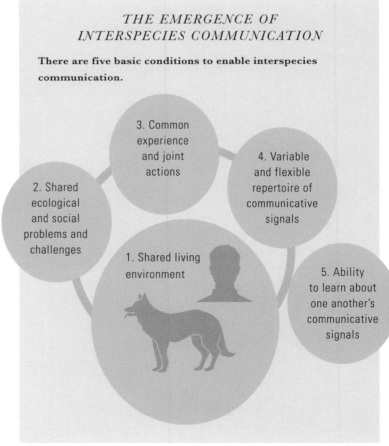

3. Common experience and joint actions

4. Variable and flexible repertoire of communicative signals

2. Shared ecological and social problems and challenges

1. Shared living environment

5. Ability to learn about one another's communicative signals

Visual Signals

Dogs use their whole body and face for visual signaling. In their interaction with humans, dogs predominantly use their species-specific signal set but they may develop new patterns of behavior with a signaling function toward their human partners. For example, dogs may employ gaze alternation to express their intentions similarly to the respective human gesture.

Below *The typical signal that a dog is ready to play includes lowered forepaws, an open mouth, and raised tail.*

Relaxed

Playful

Alert

SIGNALING TYPICAL EMOTIONAL STATES

Whole-body postures of emotional states differ tremendously, and so both a dog and a human are able to recognize the inner state of a dog from a distance:

A relaxed dog is usually approachable. Ears are up (not forward), head is high, mouth may be open slightly, tongue is exposed, there is a gentle slope in the hip, tail is down and relaxed, stance is loose, and there is no piloerection.

A playful dog invites others to play, and indicates that any rough behavior is not meant as a threat. Dogs have a specific play signal. The play-bow may be accompanied by excited barking or playful runs, attacks, and retreats. For this signal, dogs lower their front by bending their forepaws, orienting ears are up, mouth is open, tongue may be exposed, back is up, tail is up (and may be broadly wagged).

An alert dog is paying attention while assessing the situation. Ears are positioned forward (may twitch), eyes are wide, mouth is closed, tail is horizontal (and may move slowly), body shows slight forward lean.

A fearful and nervous dog shows signs of submission in order to prevent conflict. Head and body are lowered, tongue licks lip or is flicking, ears are pulled back and become limper. Yawning may also signal stress. The dog may turn the head away, but keep looking at the perceived threat showing the white of his eyes ("whale eye"). Eyebrows are arched and the tail is lowered or between the hind limbs (and may wag slightly). First paw may be raised and the dog may sweat through pads. Stress can also cause excessive salivation and shaking.

A fearful and aggressive dog is frightened but may attack. Head and body are lowered, pupils are dilated, ears are back, nose is wrinkled, lips are slightly curled (teeth may be visible), the fur is piloerected, and the tail is tucked.

An assertive and angry dog acts aggressively if challenged. Ears are raised, the fur is piloerected, forehead and nose may show wrinkles, the teeth are shown, nostrils and pupils are widened, mouth is open, corner of mouth is forward, the tail is upright, bristled, and may vibrate, the hind limbs are stretched out, body is leaning forward.

A submissive dog communicates to the opponent that he accepts his lower status, and aims to avoid physical confrontation. The dog makes itself look smaller, may roll on its back, turns its head to avoid eye contact, closes the eyes partly, ears are flat and back, corner of mouth is back, it flexes its limbs and feet, exposes its belly, tucks in its tail, and may sprinkle a few drops of urine.

Fearful

Fearful and aggressive

Assertive and angry

Submissive

BREED-SPECIFIC FEATURES IN VISUAL SIGNALING

Adult wolves have at least 60 different facial expressions. Several facial features (head and snout position; shape of the mouth corner, lip, forehead, eye region; ear orientation and movement) are involved in signaling. Facial gestures allow the display of very finely tuned signals.

Face morphology differs between dog breeds, and selective breeding has resulted in the loss of specific morphological features of the head and face in many breeds. This reduced the number of possible facial expressions and thus their capacity to function as visual signals. Thus, dogs with hanging ears, wrinkled skin, large eyes, or short snout do not have the potential to display specific facial signals. This may set back the recognition of visual signals both within dogs and between dogs and people.

HOW TO GREET AN UNFAMILIAR DOG

- Dogs should never be petted without the owner's permission, and it is important to let the dog make the first contact.

- One should stand straight up or crouch down while looking slightly away to avoid eye contact. This allows the dog to decide how to approach the person at his chosen speed.

- Once the dog is relaxed, it is possible to calmly pet him under the chin, neck, and the front half of the body. Offering treats to take out of the hand, while looking away, may also help.

- The next step is to gradually get the dog used to the stranger in different positions. Leaning over the dog or grabbing and hugging him should be avoided.

- Lots of dogs may feel threatened by too quick an approach and the close proximity of strangers. Some will freeze; others will bark or growl. As these reactions successfully discourage the greeting strangers, dogs learn that aggressive display keeps them away.

Left *The visual signals that a dog uses depend to a large extent on their face morphology, which varies across breeds.*

TAILS & EARS

Apart from playing a role in balance and hearing, respectively, the tail and the ears are essential components of dogs' visual communication system. The tail can be wagged with different frequency and intensity and can be held in a wide range of positions. Tails often have a distinctly colored tip that accentuates the tail's position and helps in reducing or increasing the perceived size of the dog. A wagged tail elicits a faster approach from conspecifics than a nonmoving tail.

Moveable ears also have an important communicative function. Their position and shape convey information about the assertiveness or alertness of the dog. Thus, the lack of either organs would compromise the dog's potential to transmit visual signals.

Below *The position and movement of a dog's tail can both be used to send messages to humans and other dogs.*

Below right *Dogs move their ears depending on their emotional state. Raised ears can be a sign of a playful or alert state.*

Acoustic Communication ～𝒬

Dogs belong to a relatively vocal family: Wild Canids as well as domestic dogs emit more than a dozen different types of vocalizations, which can be combined in different sequences. Many of these—including howling, barking, growling, and whining—can be found in the vocabularies of both dogs and wolves.

HOW DOG VOCALIZATION IS PRODUCED

Dogs produce their vocalizations in a similar way to the majority of mammalian acoustic signals. The required air is provided by the lungs (during exhaling), and this air first reaches the so-called "source" of the sound, the vocal folds at the dog's larynx. During vocalization the vocal folds are pulled in the way of the air and, on closing together, stop the airflow. If the pressure of the exhaled air reaches a particular threshold that can force the folds away from one other, they start to open and close cyclically as the air flows out between them, and sound is produced. While the volume of the sound is affected mainly by the air pressure, vocal fold tension (and thickness) affects the pitch. The vocal tract positioned above the larynx is also called the "filter" in terms of vocal production. The length and inner shape of the vocal tract modifies the primary sound generated by the vocal folds, by filtering out certain frequencies and adding the timbre of the final call.

Below *In dog vocalization, the sound is produced by the vocal folds, and then the vocal tract between the larynx and mouth modifies the vocalizations.*

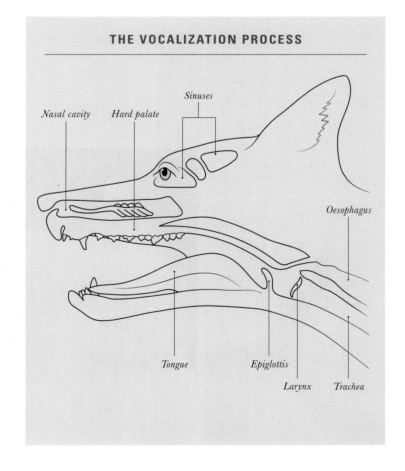

THE VOCALIZATION PROCESS

Nasal cavity

Hard palate

Sinuses

Oesophagus

Tongue

Epiglottis

Larynx

Trachea

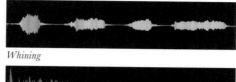

Whining

Howling

Far left *Puppies at weaning age are already capable of emitting the dog's full vocal repertoire.*

Left *Sonograms allow us to see the remarkable differences between various vocalizations.*

THE DEVELOPMENT OF VOCALIZATION IN DOGS

Dogs start emitting vocalizations right after they are born; however, their vocal repertoire is rather limited during the first 2–3 weeks. It consists mainly of different kinds of whines, screams, and squeals that are emitted due to pain, cold, hunger, or separation from the litter.

Dramatic changes emerge in the puppies' vocal behavior as soon as the eyes (days 11–14) and especially the ears (days 20–22) become operational. Vocalizations such as growling and barking are first emitted in this period, indicating that 3- to 4-week-old puppies already use vocalizations for purposes other than signaling stress. At 7–10 weeks, when a young dog becomes independent of its mother, it is already capable of emitting the full repertoire of an adult's vocalizations, including howling.

BREED DIFFERENCES IN VOCAL SIGNALING

Dog breeds have been selected by humans for different tasks on the basis of particular behavioral and anatomical peculiarities, including specific vocal habits in some cases. Some working dog types are prized for their typical vocalizations, such as the baying sound of (blood)hounds—an easily recognizable bark–howl combination signal broadcasting that the hound has found the scent of the game. Other genetic variations in the vocal habits of dog breeds may just be the product of chance. Breeds showing the closest genetic link with wolves (such as huskies) howl more frequently than less wolflike breeds.

Right *Some dog vocalizations show breed-specificity, such as bloodhounds' baying as they catch the smell of the game.*

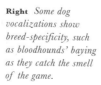

BARKING

The most striking difference between the vocal repertoires of wolves and dogs appears to be the predominant habit of barking in the dog. While wolves emit short bouts or single barks, mostly at a young age and during agonistic encounters, most companion dogs are known to be "barkers" and there are several contexts where dogs bark rather abundantly.

One theory about the evolutionary origin and function of this typical dog vocalization claims that the contagious barking of neighborhood dogs upon the arrival of an intruder (the mail carrier, for example, or somebody with a dog on a leash walking along the street) is similar to the harassing of predators in species such as corvid birds.

Another hypothesis suggests that, since domestication, dogs have shared their social space with humans, and this coexistence paved the way to new communicative interactions, including vocal signaling. Thus, barks became the type of vocalization through which dogs could convey several kinds of messages toward their human audience.

The highly variable and repetitive barks of dogs show a much broader acoustic range than wolf barks and, according to recent experimental data, humans can attribute accurate contextual and affective meaning to dog barks. However, barking is not a solely human-directed vocalization: Other dogs can also decipher information about the barking individual's identity and emotions by listening to its bark.

Below left *The barking of an aggressive dog is deep-pitched and fast— these vocalizations are easily recognized even by untrained human listeners.*

Below *Dogs also bark during play. This feature is a new addition to their vocal repertoire, as wild-living canids, such as wolves, jackals, and foxes, do not bark while playing.*

Small dog barking

Large dog barking

Right *Dog vocalizations are usually meaningful for humans. We can often decipher both the context and the inner state of a dog while listening to its barks or growls.*

THE COMMON CHARACTERISTICS OF DOG & HUMAN VOCALIZATIONS

Humans, even eight- to ten-year-old children, can judge a dog's emotions from its vocalization. This is because acoustic signals that encode basic inner mental states show considerable uniformity across species and even taxa:

1. Acoustic qualities of a vocalization are rather rigid fingerprints of the physical parameters of the body, the vocal folds, and the vocal tract. Thus, larger dogs are able to emit deeper vocalizations with a timbre that makes them sound large.

2. Evolutionary ritualization processes facilitated the emergence of vocal signals that correspond to specific predictive morphological features. For example, larger individuals have a greater chance of winning a fight, so deeper vocalizations are utilized as signals for indicating more aggressive tendencies.

3. Similar neural activation during different inner states controls the vocal production in similar ways in mammals. Therefore, an aggressive or submissive individual emits similar sounds that signal its intentions regardless of species (within reasonable limitations).

THE MEANING OF DOG VOCALIZATIONS

Disregarding the elements such as syntax, symbolism, and size of vocabulary that hallmark human language, dog vocalizations seemingly lack one other important feature that makes human conversations so meaningful—the referentiality. Dogs do not vocalize about things that are independent of their own inner state or qualities. In principle, the acoustic signals of a dog indicate its internal mental state and its indexical attributes (size, age, sex, identity). These are equally informative for humans and other dogs as well.

Olfactory Signals

It is often believed that dogs live in a world of odors, but whether they rely mainly on the sense of smell, rather than on other senses, depends on the context. For example, if clear visual cues are available, there might not be the need to rely on olfactory cues. Nevertheless, olfactory information is important for kin and individual recognition. Very young puppies (28–35 days) are able to recognize their own bed from a stranger's bed, based exclusively on olfactory information.

Pheromones are specific odorous molecules that are produced by glands in the skin and have a signaling function in communication.

SEX PHEROMONES IN DOGS

Sex pheromones are produced by the female during the estrus period. These substances originate from the vagina, urine, anal sacs, feces, and other organs. A component of this pheromone, methyl-*p*-hydroxybenzoate, has the potential to elicit mounting behavior in male dogs, but more recent observations have cast some doubt over whether this chemical is the main factor in dogs' sex pheromones. Dogs of both sexes are able to detect and discriminate these pheromones from other odors. Male dogs prefer female odors over the odors of other males in general, but their preference is even greater for the odor of a female in estrus.

The urine of male dogs most likely also contains odorous molecules that inform other males about their sex and status, but so far no specific data are available. Male dogs can discriminate their own urine from the urine of other males, and they will often urinate over that left by other males.

Below *Male dogs receive information about the reproductive status of the female by sniffing her genital region.*

THE MAMMARY PHEROMONE

During lactation, the mother dog produces the mammary pheromone, also called the appeasing pheromone. This is a mixture of fatty acids released by the sebaceous gland located in the intermammary sulcus. This pheromone is thought to help the blind puppies in finding the teat and to assist in calming the litter, although at present its role is not fully known.

A synthetic analog of the mammary pheromone seems to have some effects in reducing stress and arousal in dogs. In puppies, it may facilitate interactions with other conspecifics (during puppy classes, for example) by reducing excitability and fear. A calming effect was also reported in adult dogs in stressful situations.

However, there is a large variation in the effectiveness of the synthetic analog of the mammary pheromone, depending on the age of the dog and on the breed.

The biological mechanism of this pheromone in adult dogs is not fully known at present. As it is produced by the mother during the suckling period, it is probable that it has greater effects on puppies, rather than on adult dogs.

INTERSPECIFIC OLFACTORY SIGNALS

The pheromones in mammals share some common features, so dogs may be able to detect the estrus status in women (just as in bitches) and may be able to use sex-specific olfactory cues to discriminate between female and male humans. With human infants, dogs investigate preferentially certain areas of their body—the face and upper limbs. This may indicate that these body parts produce distinctive odors in humans, providing specific information for the dogs; or that odors in these areas are more perceptible.

Left *During lactation, the mother dog releases the mammary pheromone, which is thought to help calm the litter.*

CHAPTER 4

Sensing, Thinking
& Personality

How Dogs Think ⟿

Despite the fact that most of us have a good lay sense of what constitutes "thinking," defining the term scientifically and adapting it to animals is far from straightforward. Thinking generally refers to complex mental (cognitive) processes that involve the combination and organization of multiple—genetically influenced or learned—concepts, ideas, or percepts (so-called mental representations) in the service of arriving at a goal. Thus, the concept of thinking is inherently intertwined with the concept of problem-solving. Thinking allows the dog to make predictions about the events in the environment, contributing greatly to its chances of survival.

DEVELOPMENT OF SPECIES-SPECIFIC COGNITIVE ABILITIES

The dog's cognitive abilities are shaped by both its physical and social environment. Regarding the latter, the dog represents a unique case as its social environment consists of both conspecifics and humans. Consequently, while the cognitive processes of dogs (as one of the canine species) have been adapted to living and hunting in packs, domestication has also favored the development of skills fostering cooperation with humans. The former facilitated the evolution of skills, such as calculating the motion path of

Below *Like their closest relatives, the gray wolves, dogs' cognitive skills include those adapted to living in packs.*

conspecifics and synchronizing behavior with pack mates, while the latter led to the development of an ability to interpret the communicative signals of humans. The superior social skills allow the dog to overcome its own cognitive constraints by learning from others.

Dogs perform poorly in situations that require them first to distance themselves from a piece of visible but inaccessible food (such as when made to detour around a fence) because their ancestor is a flatland carnivore for whom moving away from the prey in their primary physical environment is an inefficient solution to predation problems. Importantly, however, dogs' performance improves significantly after watching another dog or a human present the solution for a detour problem.

BREED-SPECIFIC DIFFERENCES IN PROBLEM-SOLVING ABILITIES

Next to the general selection of the human environment, within-species variations are highly relevant when considering dogs' problem-solving abilities. The reason for this is that different breeds have been selected for assisting humans in different tasks, and consequently breeds have been traditionally classified by the specific problems they are most adept at solving (such as hunting or shepherding).

Dogs that have been selected to perform tasks that require constant cooperation with a human (such as herding dog breeds) are more proficient in reading the communicative signals of humans. For example, these dogs seem to be more skilled at following the pointing

gesture of a human to locate a hidden object than breeds that have been selected to perform tasks individually, such as hounds. Importantly, the same ability is also influenced by simple morphological differences between breeds. Namely, dogs with short skulls (brachycephalic), such as boxers, tend to be more successful in this kind of situation than dogs with long heads (dolichocephalic), such as greyhounds, because the position of the eyes makes it easier for them to focus their attention forward and not be distracted by the periphery.

Below One of the highly variable characteristics of dog breeds is the shape of the skull. Research provides evidence that brachycephalic and dolichocephalic dogs differ in their visual communication skills.

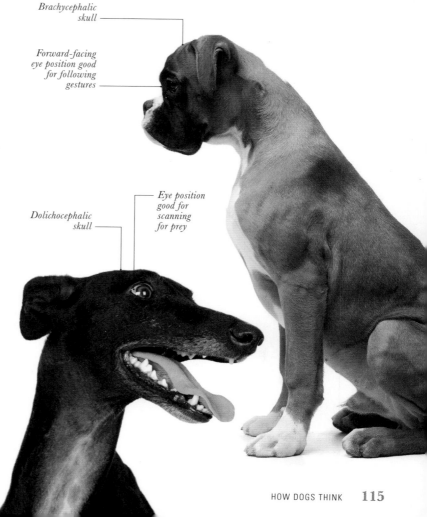

Brachycephalic skull

Forward-facing eye position good for following gestures

Dolichocephalic skull

Eye position good for scanning for prey

FUNDAMENTAL MECHANISMS OF COGNITIVE PROCESSING

Thinking can be viewed as a collection of mental mechanisms that deal with environmental information received by sensory organs, build up complex hierarchies of representations by multilayered learning processes, and are used to make decisions about the most optimal action to be carried out.

Perception and attention

Although dogs' processes of perception and attention are in general similar to those found in other mammals, their mental processes diverge in the way they allocate attention to different elements of the environment.

One key difference is a marked focus on social stimuli, with a tendency to direct attention to humans. Moreover, dogs also differentiate in the amount of attention they allocate to people based on their relationship with the person, focusing more on people with whom they have a close relationship.

Learning

Dogs are able to obtain information in several ways. So-called *classical conditioning* makes them able to detect connections between different elements of the environment and to make predictions about future events. For example, dogs quickly learn that the sound of the cupboard opening means that it is time for dinner and respond to that sound with increased secretion of saliva. Dogs are also able to connect their own behavior with consequences, such as rewards or punishments (*operant conditioning*), which makes it possible for humans to teach their dogs—for example—not to take food from the dining table.

In other instances, dogs are able to modify their behavior after having observed the behavior of another individual, either human or another dog. This is called *social learning* and allows for a more flexible way of knowledge acquisition.

Memory

Dogs are also capable of storing the learned information in memory and recalling it at a later point. This refers both to remembering associations between elements of the environment (the sound of the cupboard opening and feeding) and remembering learned actions, such as opening the lid of a box with their paws. Memory, especially

Right *The mirror test is widely used to measure self-awareness in animals. Dogs show little interest in looking at their reflection in mirrors. This may suggest that they have a limited capacity for self-reflection.*

Below *Although olfaction is generally believed to be dogs' most powerful and important sense, dogs readily learn from humans using the visual sensory channel and their visual attention is a reliable indicator of their mental activity.*

MODERN TECHNOLOGY IN THE SERVICE OF MEASURING COGNITIVE ABILITIES IN DOGS

Modern technology has equipped researchers with ingenious noninvasive tools that provide a window onto the neural processing underlying dogs' thinking.

Eye-tracking Looking behavior is a good measure for mental activities. Based on the direction of a dog's gaze, we can make inferences about its preferences and interpretation of a scenario.

EEG By measuring the electric activity of the brain by the means of electrodes placed on the skull (below), researchers collect information about how different types of stimulation facilitate neural processing.

fMRI Totally noninvasive brain-imaging methods have been successfully used with dogs in the past years to uncover brain areas related to specific cognitive skills.

remembering for a long time, is a key factor in taking decisions that can be made on the basis of a much broader experience. Dogs' memory depends on the circumstances of establishment and also the content. They may remember their owners for many years—although no objective evidence is available.

Complex cognitive processes in dogs

Self-awareness refers to the capacity to step outside of the egocentric perspective and form a mental representation about the self. Investigating self-awareness in a nonverbal species is extremely challenging. However, to date, there are no convincing experimental data or case studies that would suggest that dogs possess these abilities. Although it seems that dogs are unable to recognize themselves in a mirror, their self-awareness may have a different nature.

Understanding & Learning about the Physical World ⁓℮

In order to survive, canines need to be able to navigate in their environment and find food. Dogs are obviously not an exception, even though those living in the anthropogenic environment will be under the care of humans. In any case, such abilities are based on some forms of mental representation of the physical environment, which develops on the basis of genetic predispositions and developmental experience.

OBJECT PERMANENCE

Hunting for prey requires the ability to detect and follow it, even when it temporarily disappears from view, for example, behind trees or rocks, during the chase. The ability to form and hold a mental representation of an object that is temporarily not visible is called object permanence. The ability of dogs to rapidly recover an object that was hidden

Left *Dogs, like other predators, can detect and follow an object even when it temporarily disappears from view, such as behind the trees in a forest.*

a short time before is taken as evidence that they form a mental representation of the object and of its location and retain it in their memory. Dogs probably follow a simple rule for solving this problem: They associate certain environmental events with where the object is hidden. Thus, they may learn to go to the place where they have seen the object disappear.

OBJECT FOLLOWING

A successful predator should be able to calculate the path of escaping prey. Dogs can also follow the movement trajectories of falling or flying objects. For example, they are good at predicting where a ball or a flying disc thrown for them will fall. To do this, they may use a computation process that is similar to that used by baseball players.

SPATIAL NAVIGATION

Orienting in space is fundamental for species that need to navigate in a territory to find food and other resources. Although there are several famous anecdotes of dogs finding their way home and traveling very long distances to do so, this ability might be overestimated, because these anecdotes do not take into account the number of dogs that get lost and never find their way back home.

Right *The ability to predict trajectories is important for predators hunting fast-moving prey. Dogs are skilled at predicting the movement of flying and falling objects.*

STRATEGIES OF SPATIAL NAVIGATION

Two different main strategies can be used when navigating in space.

Egocentric strategies are based on the relationship between one's own body and the environmental information (such as turning left to locate the hidden ball). This strategy is especially useful in the absence of environmental cues for orientation. During the hunt, a predator chasing a prey may not pay much attention to the surroundings, and so this egocentric strategy might help in finding its way back home.

Dogs can also orient themselves in the absence of environmental cues by integrating information acquired by judging the distance traveled on foot, the speed, and the directional changes (*path integration*). In an experiment, dogs were led with covered eyes and ears along an L-shaped route in a large space. Once released, they were able to make the correct turn and run toward the target at the starting location. They could also judge the distance traveled and search for the target place.

Allocentric strategies are based on the spatial relationship between various objects in the space (the hidden ball is between the chair and the table, for example). Dogs rely both on *beacons*, which are direct visual markers in the vicinity of the hidden target, and *landmarks*, which are larger, visually outstanding objects in the environment that indirectly provide information about the location of the target.

In practice, dogs rely on both allocentric and egocentric strategies when navigating. For spatial orientation, dogs utilize visual, auditory, and olfactory cues that often complement one other. Experiments performed in the laboratory suggest that, on a small scale, dogs prefer to use egocentric strategies for navigation and resort to allocentric strategies only if the direct visual relationship between the dog and the target is disrupted and in cases where they are orienting in a complex environment.

Dogs' flexible use of different strategies might be even more task and context dependent than previously thought. For example, when acquiring spatial information socially from a human demonstrator in imitation tasks, dogs rely preferentially on allocentric strategies to locate the object on which the demonstrator acted. However, this tendency may be caused by the social nature of the task.

Below and below right
Observing dogs in activities where they are required to manipulate or find objects helps us to learn more about how they understand the physical world and the strategies they employ.

MANIPULATING OBJECTS: PULLING STRINGS

While humans excel in manipulating objects, our canine partners lag behind in such skills, although they also need to rely on some handling abilities, for example when dismantling prey. This difference also makes it likely that dogs' mental representation of objects and their features differs significantly from that of humans.

Despite this difference, it is worth investigating how much canines may understand from a task in which they can obtain a small reward by pulling a string. Typically, family dogs learn to pull strings, independently of their orientation, if a treat is attached to the opposite end. However, when the task is made more difficult, for example by putting down two crossed strings with only one having a treat at the end, dogs do not seem to choose the correct string to pull. Thus, they do not seem to grasp the idea that things need to be connected physically for them to move together. Interestingly, wolves make very similar errors when confronted with such tasks, so domestication did not cause this skill to deteriorate.

DOGS WATCHING TV

Dogs can utilize real-size video-projected images of the hiding event to locate an object in the real world, at least if the video projection is watched in the same room where the object is actually hidden. Thus, dogs are able to rely on representations of images that are watched on a screen to solve problems in their environment. It is not known, however, whether they are able to understand the relationship between reality and fiction.

DISCRIMINATION OF AMOUNTS

Dogs can discriminate between quantities and they do so using a mechanism that relies on some perceptual features that differentiate big from small. For example, a larger amount of food evokes a bigger visual image and thus is preferred over the smaller amount. Similarly to other species, dogs are more likely to show a preference for the bigger amount if the difference is large. If the two amounts get closer quantitatively, dogs make more and more errors. Neither wolves nor dogs are able to choose the bigger if the ratio between the two amounts is 3:4.

Understanding & Learning about the Social World ～〇

Dogs have been exposed to selection that favored the development of an understanding of the social world. Living in the anthropogenic environment, dogs must be able to acquire and store information coming from a range of social partners in order to work well among humans—a phenomenon called *social learning*.

SOCIAL LEARNING VERSUS INDIVIDUAL LEARNING

Social learning is the dog's ability to make use of others' knowledge in acquiring information about the world. This ability has been suggested as laying the groundwork for complex mental processing as it widens the scope of information that can potentially be accessed. An individual learner is bound to go through timely—and sometimes dangerous—trial-and-error learning, and it is also constrained by species-specific evolved mechanisms that necessarily bias its thinking. In contrast, a social learner may acquire knowledge in a more flexible way by observing and copying others.

WHAT A DOG CAN LEARN VIA SOCIAL INTERACTIONS

For most species, social learning entails interactions between conspecifics. However, dogs represent a special case among animals as they are not only adept at learning from other dogs, but are also capable of using information presented to them by humans. Thus, we may differentiate between *intraspecific* and *interspecific* learning.

Below *Due to their special domestication history, dogs tend to pay increased social attention to humans and in many situations they pay somewhat less attention to their conspecifics than wolves do.*

INTRASPECIFIC LEARNING— INFORMATION TRANSFER FROM DOG TO DOG

Even though they are predators, dogs are able to learn socially about food, which affects their preference. Alongside genetically influenced preferences or disgust toward certain flavors, dogs can also follow the example of conspecifics in deciding what to eat. Dog embryos in the womb experience the mother's diet (via the joint blood circulation) and as puppies when sucking her milk. Older dogs may sniff the breath of their dog (or even human) companion, and this can make them show a preference for what the other has just consumed at a later time.

Dogs may also use different mechanisms, such as direct observation, when relying on conspecifics' examples in overcoming problems. In studies exploring how observation leads to learning and

Left *In this activity to test dogs' learning, the dog must make the connection between an action and a reward, pulling the rolling board under the fence to get the reward.*

Below *Dogs gain information about each other by smelling the other's face. Apart from recognizing the identity of a partner, dogs may also learn about what the other has just eaten.*

knowing, a selected dog (the demonstrator) is trained to perform a task, such as using its paws to pull a tray containing food inside its cage. Following that, other, task-naïve dogs are allowed to observe the demonstrator dog solving the problem. Next, one of the observer dogs is confronted with the task to see how much it grasped by watching.

Results show that dogs have a tendency to reproduce the observed actions, and thus find the solution easier than by individual learning, relying on their trial-error skills. By doing so they may rely on different kinds of information. For example, it may be that the behavior of the demonstrator dog directs the observer's attention to certain parts of the object or the environment and later this helps the learner to figure out the solution on its own. However, dogs may also be capable of recognizing the relationship between the demonstrator's goal and action. In this case, the observer dogs may choose to act in the same way as they saw the demonstrator act.

INTERSPECIFIC LEARNING—INFORMATION TRANSFER FROM HUMAN TO DOG

Research has shown that dogs are very skilled at learning from humans. An inherent problem with learning from people is that they behave and execute actions quite differently from dogs. Dogs may simply attend to the overall feature of the action—for example, what part of an object is touched by the demonstrator—but more mental power is needed to copy the behavior of the human because the dog may need to learn how he can carry out a human's actions with his different body shape and motor abilities.

For example, dogs may learn to use their paws for opening a container by observing another dog perform the action, but can they do the same after watching a human demonstrator use their hands for the same purpose? Experiments suggest that dogs are capable of this. After being taught to perform certain actions on command, they are also able to imitate novel behaviors on hearing the command "Do as I do." Importantly, dogs do not exactly reproduce the observed actions, but choose a functionally appropriate way from their own action repertoire to arrive at the same goal. For example, after watching a human pull an object out of a container with her hand, the dog can also obtain the object but would use its mouth for this purpose.

Humans take an active role in learning scenarios. This involves communicating their intentions to pass on information with the help of signals, such as making eye contact, calling out to the dog, and using a special voice pitch. Dogs, similarly to children, are quite proficient in reading

these signals and increase their attention to information that is presented in this manner. In the detour task, dogs learn the solution of going around the fence faster if the human not only demonstrates the action, but also addresses the dog beforehand. Importantly, while dogs readily acquire new behavioral patterns in interspecific contexts, there is reason to assume the dog is showing obedience to instruction rather than actual learning. That is, dogs respond to the commands of the human, but perform quite poorly in utilizing the knowledge in novel contexts.

THE ROLE OF ATTENTION IN EFFICIENT SOCIAL LEARNING

Learning from and about social partners requires first and foremost a predisposition to direct attention to them. Dogs living in the human family are especially prone to focus their attention on humans, sometimes even more so than on their conspecifics. This skill emerges early in puppies when they spontaneously learn to look at the head and face of humans. During further development dogs also acquire a unique tolerance and preference for eye contact. This is especially noteworthy considering that most animal species interpret sustained eye contact as a threat, yet for dogs it seems an important tool in interspecies communication.

Dogs can also judge the direction of human attention. Dogs are more likely to beg for food from a person whose head is turned in their direction or whose eyes are open, rather than from someone with closed eyes or head turned away.

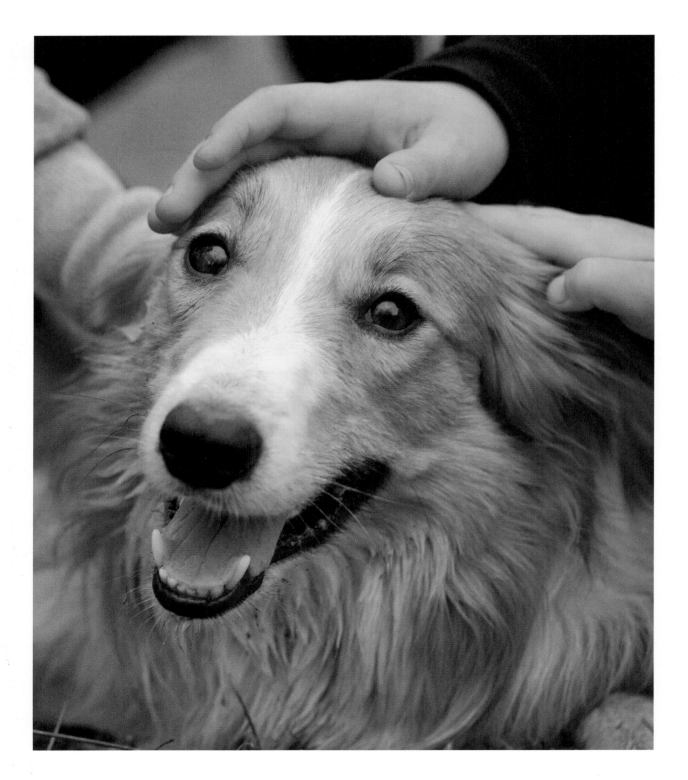

SEEKING INFORMATION

Dogs' participation in interspecific learning scenarios is not constrained to paying attention to the demonstrations or communicative signals of the human partner, but also extends to actively seeking information from them. When faced with a novel stimulus, especially if it is viewed as potentially threatening (such as seeing an electric fan in the room), dogs tend to stop and look back at their owners before initiating any contact. The reaction of the owner often determines what the dog's reaction will be—if the owner shows signs of fear, the dog will also refrain from making contact. This phenomenon, called social referencing, not only shows that dogs actively seek information from humans but also illustrates that dogs are tuned to decode the emotional reactions of the human.

SOCIAL COMPETENCE

Dogs growing up with a human family acquire a set of social skills that help them to form a close social relationship with their owner and other human companions. The overall performance—in other words, fitting into the family—has been described as the dogs' social competence. Genetic predisposition, the breed, and the quality of social experience may modify the level of social competence in dogs. But, in general, most dogs are successful in getting over these hurdles based on their skills of learning from humans, relying on human communicative signals, learning and following rules, and participating in cooperative interactions.

WORD LEARNING

While word learning is only part of the complex ability that is exhibited in humans' speech production and perception, it in itself poses a great challenge to the cognitive system. In the case of dogs, two famous instances of word learning have been studied and tested in detail. It was claimed that a border collie, Rico, knew the meanings of 200 words, as shown by his ability to fetch the objects upon hearing their names. Chaser, another border collie, was reported to fetch more than 1,000 objects upon hearing their names uttered by the experimenter. She was also able to learn a few labels referring to categories of objects.

Chaser's and Rico's skills are noteworthy in themselves, and also in regard to the mechanism of learning. In some specific experimental trials, the experimenter placed a number of objects inside a room, of which only one was unfamiliar to the dog. Then, the dog was asked to fetch an object labeled by a novel word. Both Rico and Chaser could "infer" that the unfamiliar word may refer to the unfamiliar object and made the correct choice even without having prior knowledge of the meaning of the word in question. It is important to note, however, that both dogs went through years of extensive training to reach this level of performance.

Left and below *Being part of a human family allows dogs to learn interaction and the social skills required for a range of situations.*

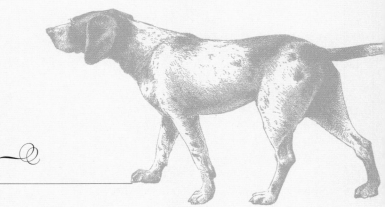

Canine Personality ~

At the beginning of the twentieth century, Ivan Pavlov, the Russian physiologist, studied the individual differences in the dogs that participated in his experiments on associative learning. He noticed that individual dogs reacted differently to handling and to the stimuli to be learned.

EARLY IDEAS ABOUT CANINE PERSONALITIES

Based on the four categories set up by Hippocrates, the ancient Greek physician, he divided dogs into four different types. *Choleric* dogs were active, unbalanced, and tended to be aggressive; *phlegmatic* dogs were quiet, slow, restrained, and persistent; *sanguine* dogs were reactive to novel stimuli, mobile, and sleepy from monotony; *melancholic* dogs were nervous, inhibited, and struggled when restrained.

These categories were adapted later by many dog trainers, but current personality research in canines does not favor such typology. Instead, researchers usually rely on methods used in human personality investigations, which emphasize that every individual occupies a particular place within the dimensions of personality.

BREED-RELATED DIFFERENCES

Breeds do not have "personality" but personality traits can be used to compare different breeds. Herding dogs and cooperative hunting dogs (such as pointers and retrievers) were found to be highly trainable, while toy dogs and hounds were reported to be less trainable. Terriers were described as being bolder than hounds and herding dogs, and other studies characterized them also as energetic, excitable, and reactive. Selecting for shows or work are also important factors that have a divergent effect on the behavior traits both within and between dog breeds.

However, in spite of breeding purebred dogs, acquiring one does not perfectly guarantee a breed-characteristic behavior pattern. Much within breed variation is caused by environmental factors, experience, and the interplay between genes and environment.

THE GENERAL STRUCTURE OF CANINE PERSONALITY

Personality traits are defined as an individual's complex behavioral patterns that are stable across time and context. The establishment of a personality trait depends crucially on how it is measured. Utilizing a standardized behavior test battery is an objective way to obtain such information, although some personality traits, such as aggression, are difficult to measure in a reliable way. Asking dog owners to fill in questionnaires may result in more subjective assessment, but their answers may correspond better to the dog's everyday behavior.

The review of many studies led researchers to identify six main personality traits in dogs. It remains to be seen whether these general categories of dog personality become standard.

Fearfulness (opposite of nervous stability, self-confidence, boldness, courage) is related to the approach/avoidance of novel objects, and activity/reactivity/excitability in novel situations.

Sociability (synonymous to extroversion) refers to the tendency to initiate friendly interactions with familiar and unfamiliar people or other dogs.

Responsiveness to training (synonymous to problem solving, willingness to work, trainability, and cooperation) indicates how the dog cooperates when working with people, whether it learns quickly in new situations, and how playful it is.

Fearful

Sociable

Aggressive

Assertive

Responsive to training

Aggression reflects the inclination to solve social disputes by biting, growling, and snapping at people or other dogs. Aggressive behavior is sometimes divided into subcategories on the basis of the causation (for example, protecting territory) and the social target (such as a child).

Assertiveness (opposite to submissiveness, withdrawnness) is reflected in such behaviors as refusing to move out of a person's path, or leading when walking in a group.

Activity refers to a preference to display locomotor activity in familiar or novel open areas.

There is some debate about whether *assertiveness* is an independent trait from *aggression* and whether *activity* should be considered as a personality trait.

Above *The personalities of individual dogs tend to be characterized by one of 5–7 broad behavioral traits.*

Below *Aggression and assertiveness are the most consistent traits in puppies. Personality traits in adult dogs are relatively stable over time.*

GENETIC FACTORS IN THE DEVELOPMENT OF PERSONALITY TRAITS

So far, numerous genes have been mapped as being responsible for morphological traits and diseases, but the identification of genes underlying personality traits is still lagging behind. Although many specific genes related to brain functions (including androgen, serotonin, dopamine, and other systems) have already been sequenced, personality traits are controlled by many (polygenic) genes. Thus, each gene has a relatively small contribution to the expression of a specific personality trait, making this effect very difficult to verify. This often means that the effects of a gene surface only in specific environments. Due to this complexity, genetic testing for personality traits is currently quite unrealistic.

ENVIRONMENTAL FACTORS IN DEVELOPMENT OF PERSONALITY TRAITS

Investigating the effect of specific environmental influences on personality has not much predictive power either, because the potential number of such factors is endless, their effects are often small, and they may differ across breeds, rearing practices, and countries.

For example, Australian and German men were reported to have more disobedient, less trainable, bolder, and less sociable dogs than their countrywomen. In contrast, in a behavior test conducted in Austria, dogs belonging to men behaved more socially than dogs with female owners. Similarly, older owners living in Australia reported that their dogs were more likely to appear anxious, while in Germany young owners were found to have the least calm dogs. German owners who said their dog was more important to them than any other living being stated that their dogs were highly reactive to their emotions, but this link was not so strong in Hungary.

DIFFERENCE BETWEEN TEMPERAMENT & PERSONALITY

Temperament should not be used synonymously with personality. It is defined as early emerging behavior tendencies that are highly inheritable. *Personality* develops as the temperament traits present in young puppies are being modified by maturation and experience. Dogs' personalities stabilize around the age of 1–2 years, but they continue to change, albeit at a slower pace.

IS THERE PERSONALITY-MATCHING BETWEEN OWNERS & THEIR DOGS?

People may choose an individual dog or a breed because they find some aspects of their behavior attractive and/or similar to their own. Accordingly, the owner's breed choice could consciously or unconsciously reflect the owner's personality. Owners of "vicious breeds" scored themselves higher in sensation seeking and primary psychopathy; people with low agreeableness, high neuroticism, and high conscientiousness also preferred breeds bred for higher aggressive tendencies. Neurotic owners were more likely to assess their dogs as nervous, anxious, and emotionally less stable; extrovert owners assessed their dogs as energetic, enthusiastic, and sociable; moreover, conscientiousness, agreeableness, and openness traits were also similar between the dogs and their owners. Interestingly, single dogs showed the highest similarity to their owner, whereas in the case of multi-dog households the dogs' similarity patterns complemented each other.

Low agreeableness and sensation seeking

Anxious and worried

Energetic and agreeable

CHAPTER 5

Dogs & People

Dogs in Culture ~◯

Above *In Ancient Rome black dogs were praised as home guards as they were thought to be the most fearsome to intruders.*

Nowadays, in most human societies, dogs make everyday appearances in the cultural environment and thinking of people. Dogs excel as main characters in blockbuster movies and TV series, and individual citizens as well as communities spend astronomical amounts on improving their dogs' lives, including through education, veterinary costs, and insurance. Large numbers of people seem to be living in a dog-centric society, where the beneficial aspects of dog keeping dominate and the majority of people have a pro-canine attitude.

CANINE "PAW PRINTS" ACROSS HUMAN CULTURE

Dogs have been among the favorite subjects of painters since the first cave paintings were created thousands of years ago. Old paintings are of great importance in helping us learn both how the dogs of times bygone looked and what the relationship between dogs and

Right *The intimate relationship between dog and owner was often immortalized by artists. Wealthy clients, for example, modeled for painters with their lapdogs.*

humans was many hundreds of thousands of years ago. Dogs depicted on the wall paintings of houses in ancient Rome, or on the canvases of medieval painters, provide an excellent insight into the multitude of various breed types that existed centuries ago. These illustrations are also useful as reference for dog breeders and veterinarians nowadays, who are interested in the historical appearance of today's popular dog breeds.

Old paintings of dogs are also interesting as "windows on the past"—allowing us a glimpse into the role of dogs in the lives of our predecessors. Old paintings that feature dogs often focus on hunting scenes—testifying to the importance of dogs in the favorite sport of the aristocracy.

Dogs were also immortalized in portraits of wealthy members of society—in the form of lapdogs or other types of beloved canine companions—proving that dogs with essentially no other task than "being a pet" existed in the past just as they do today.

SHAPED BY TECHNOLOGY & SHAPING CULTURE

Hunting is one area of human culture that has been significantly shaped by dogs—and vice versa, as hunters selected the largest range of working dog breeds, including dozens of sight and scent hounds, terriers and dachshunds, retrievers, setters, and pointers.

Along with advances in weapon-making and horse-training, dogs had the heaviest impact on the development of the increasingly refined hunting methods employed by our forebears. The different clusters of hunting dogs are an excellent

example of how directional selection works when applied to behavior. The ancient kennel masters selected hounds for particular, well-chosen aspects of their original predatory behavior sequence. While, for example, sight hounds were selected mostly for their keen vision, fast and enduring gallop, and prey-subduing capacity, in gundogs such as pointers the breeders selected for their "before-attack" freezing—a good indicator for the hunter of the direction of the hidden game.

Above *Across the millennia, talented dog breeders developed doze[n] of hunting dog types, which can be regarded [as] highly specialized "tool[s]" for aiding their master[s to] subdue every possible game species.*

FOOD OR FRIEND?

Probably one of the biggest cultural differences regarding our relationship with dogs is the eating or not of dog flesh. This topic evokes extreme emotional reactions from many people, making factual consideration of the issue almost impossible. Today many in the West regard eating dog meat as the "crime" of particular countries, mostly in Asia. But they tend to forget that dog-meat butcheries were still in business (although in moderate quantities) at the beginning of the twentieth century in Europe—as photographic evidence from Belle Époque Paris shows. Thus, the ban on eating dog is much more of a cultural taboo among the ever-growing number of dog-loving citizens than a biological necessity.

WE ARE LIVING IN A DOG-LOVING SOCIETY

It is often said that the secret of a successful movie is to include children and/or dogs among the cast. The popularity of dogs as main characters in fiction started well before cinematography. It is likely that practically every country around the world has its own dog heroes in its storybooks, and the never fading success of such classic characters as Lassie or Rin Tin Tin shows that the human mind is still receptive to the positive emotional input elicited by our canine friends. Featuring dogs in cultural endeavors such as popular TV series can have a strong impact on the popularity of particular dog breeds.

CULTURAL TOLERANCE OR CONTROL OF DOGS

From simple things such as admitting dogs to restaurants to such drastic decisions as shooting dogs that harass wildlife, each country has its own system of rules and unwritten collective attitudes for handling various issues involving dogs.

One particularly strong motivation behind dog control is disease prevention. Diseases that can spread from dogs to humans do exist; probably the most infamous is rabies due to its dramatic appearance, terrible symptoms, and incurability. Concentrated mostly in Africa and Southeast Asia, where millions of pariah dogs live hopelessly far from ever being vaccinated, rabies still claims about 60,000 human deaths yearly. Other ailments, primarily parasites spreading from dogs to humans, are less obvious but nevertheless still mean danger. However, we should hope that our culture moves in the direction of emphasizing good hygiene and veterinary prevention, instead of deeming dogs as dangerous, "flea-ridden beasts."

DEPICTION OF DOGS IN CAVE ART

It is interesting that dogs are not typically portrayed in caves where drawings of hunting scenes can be found. The first depictions of dogs occur much later than one would expect, assuming that dogs were domesticated 16,000–32,000 years ago. The dogs shown in these pictures, dated 4000–6000 BCE, are participating mostly in hunts and have a curly tail. Armenian herdsmen drew images of herding dogs on the cave walls and large stones around 1000–3000 BCE.

Opposite *Romantic dog stories, such as* Lassie Come Home, *popularized the dog as "man's best friend" all around the world. Eric Knight's classic novel yielded several movie adaptations.*

Left *Compared to other animals such as the bison and horse, dogs appeared much later on cave paintings, suggesting that dogs were domesticated relatively late.*

Below *Today dog meat is definitely a taboo for most Europeans. However, at the end of the nineteenth century, specialized butchers supplied it for those who wanted to eat it.*

Attachment to Humans

There is a common impression that the dog is unique among animal domesticates in the sense that there is often an extraordinary bond between dogs and their owners. Indeed, dog–human companionship is rooted in dogs' long evolutionary history and as a result dogs became the prototype of companion animal by developing the potential to live with humans if socialized appropriately. A well-functioning attachment between dog and owner is a prerequisite for a good relationship, emotional stability, and effective cooperation.

WHAT IS ATTACHMENT?

The term "attachment" does not refer to an elusive emotional phenomenon and it cannot be simplified to overall partner preference or absence of fear from a familiar individual. Attachment is a *behavior control system*, which manifests itself as long-lasting attraction to the "object of attachment" (attachment figure) in the form of particular behaviors (for example, proximity and contact seeking). This behavior control system evokes a particular set of actions in stress situations that aim to reestablish contact with the attachment figure. Attachment behavior is typical for many animals including the domestic chick, the horse, and the dog, and manifests also in humans (babies).

Below *Rather than paying attention to each other, companion dogs prefer to be close to their human caregivers, and will look to them for security.*

It is important not to confuse attachment with dependency because they describe different aspects of a relationship. Attachment refers to physical accessibility and emotional availability of the caregiver, and it is fueled by the subject's social and mental needs. In contrast, the term "dependency" is associated with the satisfaction of basic needs such as food and shelter. Importantly, however, the two phenomena are not easy to disentangle because they often manifest themselves in similar behaviors.

DOG–HUMAN ATTACHMENT: ASSESSMENT & CLASSIFICATION

Although from the layperson's perspective the dog–owner relationship was always seen as a manifestation of attachment, objective scientific data had been lacking. Recently, dog researchers have utilized a standard procedure, the so-called Strange Situation Test, which was originally developed to assess the behavioral interactions between a (human) mother and her infant, to observe the manifestation of attachment behavior between the dog and its owner.

In these assessments dogs are observed in a new place that is thought to evoke some moderate stress in them. Most dogs seem to tolerate this situation well in the presence of their owner and also engage in playful social interaction with a strange human. However, they usually stop playing if the owner leaves, and show many signs of separation behavior. In many dogs, stress can be enhanced by leaving them alone for a short time at this strange place and cannot be alleviated by the presence of another person. These observations clearly show that the owner represents a very "special person" for a dog whose main role is to provide security.

Mother–infant affiliative relationships are often categorized according to the "secure" or "insecure" nature of the attachment, and researchers have also

Above *Showing preference for the caregiver in a stressful situation is a typical characteristic of attachment behavior shared by dogs and infants.*

Safe haven

Secure base

Separation

Greeting

Right *There are four important features of attachment behavior in dogs (see also box below).*

attempted to detect similar patterns in dogs' attachment. Although the dog–human attachment cannot be clearly categorized as secure or insecure, research suggests that the strength of the attachment relationship can be classified along a continuum according to the intensity of separation anxiety. Dogs that tend to spend much time close to the owner are often referred to as being "more attached"; however, this may rather be a sign of feeling "insecure." In some cases dogs show an extreme reaction to separation however, studies have failed to show any direct relationship between patterns of attachment behavior and separation anxiety disorders in dogs.

BEHAVIORAL HALLMARKS OF ATTACHMENT

Here are the main features of attachment (although some dogs may deviate from the typical pattern):

Secure base effect: When exploring an unfamiliar environment, the dog contacts the owner regularly.

Safe haven effect: When experiencing danger, the dog takes protection near the owner by displaying specific proximity- and contact-seeking behaviors.

Separation behavior: If the dog is left alone in an unfamiliar place, he shows signs of separation and tries to reestablish contact with the owner. Such "separation distress" cannot be effectively alleviated by a person other than the owner.

Distinctive behaviors toward the caregiver/owner: Dogs show specific greeting behaviors and signs of behavioral relaxation in the presence of, or upon being reunited with, the owner after stressful separation.

RECIPROCITY & FLEXIBILITY

The dog–human attachment relationship is bidirectional: Dogs tend to show emotional and behavioral signs of attachment toward humans, and in parallel humans readily perceive this relationship as attachment entailing the subjective impression of psychological connectedness. Individualized attachment to a human caregiver develops throughout the life of a dog and, unless drastic changes in the dogs' social relationships happen, the adult dog's attachment toward its owners is fairly stable over long periods of time.

Importantly, however, dogs do not need to be acquired in early puppyhood for an attachment to develop and even breaking the attachment relationship does not impair most dogs' ability to form new attachment relationships later in life. Adult dogs from other families or from shelters may also be able to establish strong attachment to their new human caregiver. Such flexibility of establishing new attachment relationships even at a late age is unique to domestic dogs.

EVOLUTIONARY ORIGINS & DEVELOPMENT

Although there is some disagreement over the evolutionary origin of dogs' infant-like attachment behavior, domestication has probably contributed to the emergence of this social skill. Much of the recent scientific debate is about the relative contribution of domestication (genetic predispositions) and social experiences during life (socialization) to the phenomenon. Comparative

investigations of extensively socialized wolves and dogs indicate that, despite much experience with humans, the members of the former species do not develop doglike attachment behavior toward their caregiver. Thus, the domestic dog is not a tamed wolf; multifunctional psychological relationships do exist between people and dogs.

In dogs, patterns of attachment toward humans can be observed as early as 16 weeks of age, and dog puppies show very similar behavioral patterns as those described in adult dogs. Attachment behavior in dogs may be affected by experience during development, and also by the owner's personality. Dogs showing separation-related behavior problems are more likely to belong to owners who would also describe themselves as being insecurely attached.

Above *The infant-like attachment that bonds the dog to its human caregivers is apparently lacking in wolves and thus may reflect dogs' evolutionary adaptation to the human social environment.*

Dogs as Guards, Hunters & Shepherds ⁓℮

The first useful function performed by ancient dogs is likely to be guarding. The presence of such "watchdogs" could provide a sense of security because dogs vocalized if they saw or heard something suspicious. Dingoes were mainly utilized by Aboriginal Australians as providers of warmth and watchdogs.

Naturally, this guarding behavior was not some sort of intentional service for the sake of the human partner, and there was also no goal-oriented breeding of the better individuals in this respect—humans merely benefitted from eavesdropping on dogs' intraspecific communication. However, more intentional selection during the later course of our cohabitation has significantly shaped this spontaneous vocalization.

THE FIRST WORKING DOGS: SLED DOGS

Several thousands of years ago different northern peoples developed their own sled-dog breeds and harnessing methods to suit their specific goals and environments. At the beginning of the twentieth century sled dogs played a crucial part in the exploration of the polar regions and such famous explorers as Robert Peary (North Pole) and Roald Amundsen (Antarctic expeditions) adopted the Inuit technique of using sled dogs to carry freight and passengers in extreme weather conditions and on icy terrain. Despite the introduction of motor transportation, sled dogs are still used today by some rural communities, especially in areas of Alaska and Canada.

Below *The long history of dogs being used to carry freight includes their use as sled dogs by polar explorers such as Roald Amundsen.*

SELECTION FOR COMPLEX FUNCTIONS NEEDS A LONG TIME

Though it is a widespread belief that some of the first domesticated dogs were likely bred to help humans in capturing and killing game for food, this scenario seems highly unlikely. Sophisticated cooperation during the hunt and especially the willingness to share or pass the quarry must have needed a long period of time to evolve. Evidence suggests that, even nowadays, most village dogs tend to hunt with and not for human partners (they are too noisy, chase the prey away from the hunter, etc.); without specific selection for this function, even the domesticated dog is not suitable for cooperative hunting.

FUNCTIONS OF DOGS EMERGE WHEN NEEDED

Since the dog was domesticated well before cultivation and before the domestication of other species, shepherd dogs could obviously emerge only after the emergence of domesticated herds

of goats, sheep, and cattle at human settlements, around 10,000 years ago. Prioritization of suitable individuals (or, rather, acting against the unfit ones) with the aim of developing an effective helper for the hunter or shepherd role could take a very long period of time, considering that this early, rudimentary "selection technique" was pretty far from the intentional artificial selection process we experience in today's dog-breeding practices.

Some crucial behaviors inherited from the wolf's hunting technique (such as grabbing and killing, in the case of herding; and scaring away the game and defending the prey, in the case of hunting) had to be genetically modified and controlled so that a reliable cooperative partner could be gained.

Above left *Dogs' herding behavior has evolved from wolves' predatory behaviors, but lacking the final stage of the sequence, the killing.*

Above *The Russians also took advantage of sled dogs' ability to cover harsh terrain, as seen here in Siberia in the eighteenth century.*

Dogs as Assistants ~&

The role of dogs as social companions and helpmates in the tasks of daily life dates back thousands of years, and many early civilizations regarded dogs in much the same way as we do today. Traditional ways of providing help (such as in herding or hunting), however, have gradually been supplemented with some new ways of cooperation during the last few hundred years. Dogs in modern society participate in new "coworker roles," assisting security services with drug detection, security, and rescue operations.

There are also many dogs specifically trained to help people with specific chronic diseases (such as dogs for diabetics), while others improve the lives of people living with disabilities (such as hearing dogs). Many of these new functions have important social–emotional dimensions; in addition to providing practical everyday help, these dogs also offer emotional support to their human partners.

SERVICE OR ASSISTANT DOGS

Unfortunately, there is no universally accepted system for categorizing dogs that help humans in many ways. Some dogs act mainly as sensory aids, others warn their owner of some danger, and many dogs execute actions that lie beyond the capabilities of their owners.

Guide dogs for the blind assist their blind or visually impaired owner to stop at curbs and steps, to avoid obstacles, to cross roads in a safe manner, and to provide physical and emotional support in many different situations. Dogs are trained to work by the means of a harness with a U-shaped handle that promotes direct physical connection between the dog and his human partner.

Successful teamwork involves a specific type of cooperation between partners—a flexible interchange of leadership. Namely, members of the dyad are able to adjust their behavior to the behavior of the other, and the role of initiator can vary depending on the nature of the task. The owner's role in the team is to provide directional

Below *Dogs cooperate with humans in many different ways, becoming an essential helping partner in recent decades.*

commands, while the dog's role is to react with intelligent disobedience (disobeying an unsafe or dangerous command) in order to ensure the team's safety.

Golden retrievers, Labradors, and German shepherds are most appropriate for guide-dog work, although some other breeds, including border collies, are also widely used.

Hearing dogs are trained specifically to alert their handlers who are deaf or hard of hearing to household sounds, such as a doorbell, telephone ring, alarm clock, or crying baby. The dog's role is to physically contact the owner in case of hearing some significant sound and to lead his deaf partner to the source. Nudging and touching with the paws are common behaviors that dogs offer spontaneously as an attention-getting signal, and hearing-dog training relies on using this natural inclination.

Hearing dogs must respond reliably to obedience commands from the owner. At the same time, however, they must develop a proactive attitude and a high

Above *Specifically trained dogs are increasingly employed to assist in various tasks, including searching for survivors after earthquakes.*

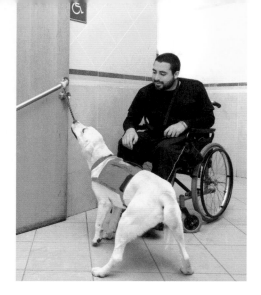

level of independence (not waiting for commands). Owner-oriented, friendly, playful, and bouncy dogs are the best candidates for the job. Most hearing dogs are retrievers or various small-breed dogs (or crossbreeds), including cocker spaniels, miniature poodles, and cockapoos.

Physical assistance dogs assist people with physical disabilities in a variety of ways. They retrieve out-of-reach objects, open and close doors, and turn lights on and off. These dogs can learn many other tasks, and after the general training they need to learn about the specific requirements of their job that depend on their owner's actual disabilities. A well-trained dog may learn to execute 50–60 actions on a given verbal or gestural signal.

These dogs need to be very quiet, should not be overexcitable, and they must not show any aggression. They should not beg from other people. While the practical help is vitally important for owners living with disabilities, these dogs also offer social–emotional benefits, eliminating feelings of fear, isolation, depression, and anxiety felt by their human partners.

Autism assistance dogs are the best-known example of service dogs for people with mental health disorders. These dogs are trained to keep their owner safe, to suppress behavioral outbursts and provide comfort when their owners are upset. They can also reduce stress for family members, helping to bring independence and a more socially inclusive life for both a child with autism and their family.

A good assistance dog is confident but not protective or overly active, well tuned to human behavior and emotions, and eager to please. Golden retrievers and Labradors are the most preferred breeds for different types of assistance dog work, but crosses such as golden retriever/ Labrador and Labrador/standard poodles and many other breeds could also be successfully trained.

Left *Physical assistance dogs are trained to perform many important tasks to assist people with disabilities.*

Right *Assistance dogs start off their training as "puppy candidates" while they are socialized to humans and other dogs and learn core skills.*

Below *A strong emotional attachment to the owner and being trained to respond to the owner's needs are essential ingredients for an assistance dog that signals (and protects) if the owner has a seizure.*

Seizure response dogs can provide help for people with epilepsy during a seizure. Their repertoire covers a variety of activities related to potential lifesaving duties. They are trained to alert family members, lie next to their owners to prevent injury, and to operate push-button devices to call the emergency services. Training of seizure response dogs differs greatly from the training of other assistant dogs. These dogs are socialized very extensively with their future owner from early on to allow the development of extreme dependence and perfect cooperation. They gradually become sensitive to small changes in human behavior and/or odor cues that precede an oncoming seizure. Some dogs may develop the ability to predict the occurrence of a seizure many minutes ahead, and alert the person in time.

MAIN STEPS TO BECOMING A SERVICE DOG

Puppy testing: Puppies around the age of 8–10 weeks are evaluated for their willingness to work, ability to retrieve objects, and eagerness to please. They are screened for friendliness, emotional stability, gentleness, and trainability.

Socialization and obedience training: Selected puppies live with foster families for about 12–18 months. They are also given the basic obedience training that all professional assistance and service dogs undergo. Experience of the natural environment and socialization to people is most important.

Specific training: For about one or two years dogs are trained to learn all the actions that may be needed to help people with disabilities.

Recipient selection and team development: Once training is complete, dog–recipient dyads are carefully matched to best suit the needs of the human and the dog.

Companion Dogs in our Cities & Apartments ~❦

Dogs kept as companions are among the most common animals that urban citizens in industrialized countries encounter day by day. Owning a dog while living in a big city is a relatively new fashion. With the exception of lapdogs and the occasional occurrence of smaller hunting dogs (such as the poodle, spaniel, and dachshund), the cities of the late nineteenth century were not teeming with the plethora of dogs we see today. There are many aspects to the debate about whether city life is good for a dog.

OWNING A CITY DOG: THE NEED TO BE TOGETHER

Many people believe that living in the city is no life for a dog. Advocates of this opinion imagine that dogs enjoy nothing more than running around in a huge area of countryside. However, dogs have evolved to prefer to be in the vicinity of humans, and so any properly socialized dog would choose to be near its owner all the time. City dogs therefore can be much happier than their kennel-kept rural

Below *Dogs make excellent companions in the city—they move just as confidently among the skyscrapers as their ancestors did in the forest.*

relatives. However, those who are planning to get a dog should think about what common activities they will do with their future canine companion. Dogs kept in a flat need to be provided with many types of activity, which means much more than simply walking side by side.

OWNING A CITY DOG: TIME SPENT ALONE

Dogs living in city flats are often left alone for most of the day while their owners are out at work or school. Many dogs do not deal well with this situation, especially when young. As puppies, they were treated to lots of fuss and attention from their owners for many weeks, and then suddenly they face the reality of being alone for long periods of time. Although there are training techniques to tackle separation-related problems in dogs, when someone knows that his/her lifestyle involves regular periods when a dog

would be left alone at home for 10–12 hours daily, a dog is probably not the best choice of pet.

DOGS & CHILDREN

It is widely believed that dogs "love children," and many feel that even for small infants a dog would be the perfect playmate. However, children under 10–12 years of age are not safe to be left alone with dogs unsupervised for a prolonged time. Dog bite injuries most commonly affect children (in Canada, for example, 85 percent of fatal dog bites involved children under 12 years). One reason for these fatalities is that young children can misinterpret the facial expressions of dogs, thinking that aggressively snarling dogs are "happy."

OWNING A DOG IS A PRIVILEGE & NOT A RIGHT

The rules and laws governing how dogs can be taken to public areas vary widely between countries and even cities. Not everybody is equally happy about dogs being around, and not every dog is necessarily willing to play immediately when another dog suddenly jumps on it. Dogs should be kept on a leash on city streets, until they reach a place where it is allowed and safe for them to run freely. Dog owners should remember that it is primarily their and their dogs' behavior that will provide the precedence—good or bad—for the next dog-related law, influencing whether that is toward more freedom or less.

Above *Most children love dogs; however, dog ownership requires a considerable amount of responsibility.*

Left *The successful coexistence of people and dogs in the city largely depends on the way dog owners introduce their pets into public contexts. A well-trained, confident, friendly dog rarely causes trouble.*

Contemporary Dog Training

Dogs, probably most dogs, learn to behave appropriately simply by being exposed to the natural environment of everyday family life. Thus, formal dog training is considered as a necessity only when the dog's spontaneous behavior is problematic for the owner. Why then is dog training so widespread?

Our accelerated and crowded lifestyle often creates challenging situations for a dog, so training might be necessary to teach the appropriate behaviors in a range of different situations (such as walking on the leash without pulling, stopping on command, and coming back to the owner if called).

Below *Dogs who attend training classes with their owners can be taught how to behave in specific situations.*

DOG TRAINING
OR DOG TEACHING?

Dog training aims to modify dogs' behavior and make it predictable. This poses the question of how dogs acquire information about their environment, process it, and modify their behavior accordingly. Most training techniques assume that dogs' behavior is mainly regulated by complex forms of associative learning. These training methods concentrate on exposing the dog to the relationship between two events (unconditioned and conditioned stimuli) and on the association between stimuli and behaviors. In contrast, cognitively oriented procedures assume that the dog's mind also deploys more sophisticated and more comprehensive processes for acquiring and processing environmental information. These views emphasize the role of attention, the social context, and the communicative interaction with the owner, and, in some cases, also take into account dogs' predisposition to acquire information socially from humans. Thus, trainers taking a cognitive approach could be said to be teaching the dog, rather than simply training it.

MOST COMMONLY USED
TRAINING METHODS

Here we summarize the four main training methods, but several variations exist. Their use varies depending on the goal. These methods differ in their effectiveness, in the speed with which the goal is achieved, and also in how they affect the dog's wellbeing. There is a growing consensus that training should not compromise the welfare of the dog.

Above *A dog focuses on its owner during training.*

Luring

The trainer lures the dog into the desired position using food or favorite toys. To make the dog lie down, the trainer holds a treat in his/her hand and, keeping the hand close to the muzzle of the dog, lowers the hand, putting it close to the floor. If the dog wants to get the food from the trainer's hand, he needs to assume a lying-down posture. Once the dog is in the desired position, the trainer gives it the treat.

Shaping

In shaping procedures, the dog's spontaneous behavior is gradually adjusted by means of appropriately timed rewards (such as food or a toy). This method involves breaking down the training goal or target behavior to be reached into simpler parts so that otherwise complex behaviors can be taught as an arrangement of simpler

parts. This way, by rewarding successive actions, the spontaneous behavior of the dog is gradually shaped to reach the final desired response. To make the dog lie down, the dog is first rewarded for spontaneously lowering its head, then for lowering the front part of its body, for stretching the front paws forward, and finally for assuming a lying-down position.

Imperative/forced

The dog is constrained and physically forced by pressure from the trainer's hand or by the collar being pulled into the desired position. To make the dog lie down, the trainer holds the leash in his/her hands and makes it pass under his/her foot. By pulling the leash, the trainer forces the dog to lie down due to the tension of the leash. Eventually, the trainer also pushes on the backbone of the dog using his/her hand. Once the dog is in the desired position, the trainer ceases all pressure.

Do as I do

This method relies on dogs' ability to imitate human actions. The dog is first taught to match his behavior to familiar actions demonstrated by his owner on the command "Do it!" This way dogs learn

an imitation rule: "Do it!" means for the dog to copy the action he has just observed. This method can be used to teach dogs novel behaviors by simply showing them. If the owner wants the dog to lie down, he or she has to show the action first, and then issue the verbal command "Do it!"

ETHICAL CONCERNS ABOUT TRAINING METHODS

In recent years a debate has emerged between trainers using so-called "positive methods" (methods that rely exclusively on rewarding desired behaviors, and avoid physically forcing or punishing the dog) and trainers using methods that include some form of physical coercion, punishment, or restraint. The use of physical constraints and punishment (especially in the case of electric collars and prong collars) poses serious ethical concerns for the dog's welfare because they may cause the dog pain and suffering. Indeed, it seems that dogs trained by physical punishment-based methods show more signs of stress during training. Dogs trained by physical punishment are also less likely to learn more novel behaviors compared to dogs trained with positive methods.

REINFORCEMENTS AND PUNISHMENTS

The most commonly used training methods involve using reinforcers and punishments to increase or decrease the frequency of appropriate and inappropriate behaviors. Today the majority of trainers prefer to work using positive reinforcements. This also means that a good trainer is very inventive and flexible in avoiding the use of punishment. For example, instead of punishing a display of undesired behavior in a specific situation (eating food from the ground, for example), dogs are taught to show an alternative appropriate action (such as go and look at the owner).

Positive punishment

Positive reinforcement

POSITIVE AND NEGATIVE TRAINING

Positive reinforcement
Stimulus or event that, when administered to the animal, increases the frequency of a specific behavior. For example, when the dog performs the desired behavior, such as lying down, the owner gives him a treat or plays with him.

Negative reinforcement
Stimulus or event that, when taken away from the animal, increases the frequency of a behavior. For example, when the dog performs the desired behavior, such as lying down, the owner stops pulling the leash.

Positive punishment
Stimulus or event that, when administered to the animal, decreases the frequency of a behavior. For example, when the dog performs an undesired behavior, such as jumping on the owner to greet him or her, the owner hits the dog's back or firmly tells the dog "No!"

Negative punishment
Stimulus or event that, when taken away from the animal, decreases the frequency of a behavior. For example, when the dog performs an undesired behavior, such as jumping on the owner to greet him or her, the owner turns his or her body in the opposite direction, thus ignoring the dog.

Dogs' Beneficial Effects on Human Health

Dogs as companion animals have become deeply embedded in human society, with an astonishing range of effects on humans, including both mental and physical health outcomes. Indeed, there is ample evidence that dogs benefit humans in many ways. In addition, dogs are effective social facilitators and adjunct therapists.

THE THERAPEUTIC VALUE OF HUMAN–DOG INTERACTIONS

Dogs were introduced as co-therapists in the 1960s, and since then they have become the most widely used species for animal-assisted interventions including psychotherapeutic, socio-educational, and socio-occupational activities.

Therapy dog is an umbrella term to describe dogs benefitting people in a therapeutic way. This includes different types of dog-assisted activities with a wide variety of potential clients. Some therapy dogs take part in visiting programs (in schools or homes for the elderly to enrich social experience). Other dogs participate in structured activities as part of a therapeutic program or practice.

Specifically trained therapy dogs can act as "social catalysts," facilitating interpersonal contact and socialization among patients as well as alleviating feelings of loneliness, reducing depression, and promoting a positive mood (especially in hospitalized children and elderly nursing-home residents).

Right *Petting is more than just a pleasant experience for dogs and owners: it has positive effects on both physical and psychological wellbeing.*

Below *Dogs may be the perfect fitness partners: they help their owners stay both mentally and physically active.*

These dogs may also serve as attention-capturing stimuli as well as positive reinforcement tools in both educational and therapeutic settings when the execution of exercises is too difficult for the learners.

Generally speaking, dog–human contact has been found to enhance the psychological and physiological parameters important to human health, wellbeing, and quality of life.

OXYTOCIN

The neurohormone oxytocin plays a significant role in the majority of the positive effects of the dog–human relationship on human health and wellbeing. Oxytocin stimulates social interaction, decreases stress hormone levels in response to social stressors, decreases the behavioral stress response (anxiolytic effects), increases the pain threshold, and boosts the immune system. Dog–human interaction has the potential to activate the oxytocinergic system in the brain of both partners. Gentle stroking or making prolonged eye contact with a companion dog produces intense emotional reactions and a substantial increase in circulating oxytocin in both dogs and humans.

RESEARCH FINDINGS ON HUMAN HEALTH & THE PSYCHOSOCIAL EFFECTS OF DOGS

Dogs keep their human partners active: Dog owners are more likely to participate regularly in physical activity.

Dogs help their human partners keep calm: Affiliative interactions (such as petting) between human and dog attenuate stress-related responses during a stressor, lowering the heart rate, blood pressure, and skin conductance, and circulating cortisol in the blood.

Dogs improve their human partners' immune system: Examination of salivary immunoglobulin levels (IgA—an indicator of good immune system functioning) indicates the beneficial short- and long-term effects of high-support-providing relationships with dogs.

Dog ownership can have a protective effect against cardiovascular risk and is associated with increased survival after suffering a heart attack. This is probably due to the more active and relaxed lifestyle of dog owners.

Dogs play a beneficial role in facilitating the psychosocial development of children: Dogs in the household contribute to healthier self-esteem and identity development in children.

Dogs promote the social inclusion of people who suffer from social isolation: The presence of a dog can function as a catalyst for interpersonal relationships, increasing positive interpersonal interactions and mood.

Dogs in the Shelter

Dog shelters were originally called dog pounds, because they were used for the temporary keeping of stray dogs caught on the streets. The early dog pounds therefore were not pleasant or humane facilities, but institutions designed to exclude potentially dangerous stray dogs from inhabited areas—secondarily also providing an opportunity for dog owners to find their lost dogs. Most of the canine inmates at these pounds faced the likelihood of being killed if not claimed by an owner.

CHANGING ROLES OF DOG SHELTERS

During the second half of the twentieth century public opinion about animal welfare and the rights of animals went through a revolutionary change, resulting in a friendlier approach to those unfortunate dogs that end up at these facilities. Today, the primary mission of dog shelters is to find the most appropriate next owner/home for any dog that is suitable for rehoming. Depending on the laws in the particular country and the domestic regulations of the shelter, euthanasia is normally only a last resort, to be applied for those dogs that are too dangerous or hopelessly sick.

WHY DO DOGS END UP AT A SHELTER?

Shelters' records show that behavioral problems that owners cannot deal with anymore are among the leading causes of relinquishment. Aggression (against humans and/or other dogs), separation anxiety, excessive barking, and difficulties with training (including being house-broken) are noted most often. Other common reasons can be allergies to dogs, local regulations against dog keeping, or changes in lifestyle.

Right *Even today, high numbers of dogs end up at shelters, where their chances of finding a new owner are often slim.*

FINDING NEW HOMES

Living at a shelter is not an ideal state for a dog because probably the most fundamental need of any socialized dog cannot be fulfilled—being attached to one particular person. Even a very short 10-minute-long dedicated face-to-face interaction between a shelter dog and a previously unknown person can lead to the initial formation of attachment. Living in a shelter must not be the fate of a dog—adoption should be the goal.

Modern, responsible dog shelters not only look for the fastest way to rehome each of their inmates, but they try to find permanent homes for them, avoiding the "boomerang-dog syndrome," when a dog enters an endless relinquishment–rehoming cycle. The key factor for successful rehoming is the complex evaluation of each dog before being offered for adoption, so that new owners have all the information they need about the dog's behavior and personality.

ALTERNATIVES TO SHELTERS

To reduce the burden on shelters, alternative solutions have recently been developed. Some charities offer help to owners in their homes, to try to resolve certain problems that might lead to rehoming. Breed-specific rescue organizations take in hand the rehoming of dogs belonging to a particular breed, usually operating with the help of volunteers, who offer temporary homes for the relinquished dogs until their adoption can be managed. The involvement of volunteer temporary owners is more and more widespread in other rescue shelters as well. This process is also advantageous for the dogs, because they are provided with accommodation and interaction that is similar to a real dog–owner relationship. Meanwhile, the volunteer can assess the dog's characteristics in order to improve the chances of successful adoption later.

Above *Shelter dogs readily form fresh attachment bonds with their new owner, even after a short initial introduction.*

Malformation in Look & Shape

There are many dog breeds that have features in their anatomy that make them unable to survive in nature in the absence of the nurturing efforts of humans. Although mixed-breed dogs can show some of these malformations as well, their typical appearance is mostly encountered in purebred dogs.

ABNORMAL SKULLS

An extremely short muzzle region of the skull (brachycephaly), as seen in the English and French bulldogs, the boxer, the Boston terrier, and the pug, causes undeniable suffering. The gradual shortening affects the upper jaw more than the lower, and a pronounced underbite develops in these breeds.

A more serious problem is the severe obstruction of the upper airways, causing almost permanent heavy respiration (noisy, choking breathing). It is therefore not recommended to let bulldogs or pugs swim, run, or sunbathe unattended due to the danger of suffocating. The shallow eye socket in pugs or bulldogs is also a consequence of a malformed skull. It is not uncommon for the eyeballs of these dogs to pop out from their sockets, causing suffering for the dog and potential blindness.

Above *Underbite is a common feature in short-skulled breeds such as the Boston terrier.*

Below *Dogs with extremely short legs but normal length backs often develop spinal malformations.*

ABNORMAL LEGS

Several breeds are characterized by their dwarfishly short legs. Dachshunds, corgis, and basset hounds all exhibit the mutation called achondroplasia, causing the normally long bones in their legs to stop growing. Interestingly, these breeds show normal development of the other bones in their skeleton, resulting in a disproportionate body structure with an extremely long back compared to their tiny legs. Dogs such as dachshunds are extremely prone to serious, incurable spine injuries and defects, the most common being intervertebral disk disease (IVDD). This condition develops when the cartilage disks between the vertebrae lose their flexibility, then herniate, and put a more or less severe pressure on the spinal cord. Dogs may suffer from pain, move less freely and willingly, or, in more serious cases, become paralyzed.

ABNORMAL SKIN

The skin of dogs has also been subject to misdirected selection. On the shar-pei the skin is so loose that it forms deep folds all over the body of the dog. Considered as "adorable" by fanciers around the world, an average "wrinkly" shar-pei goes through a large amount of suffering from a young age. One surgical intervention, almost routinely done on juvenile shar-peis is the entropion operation. Entropion is caused by the too heavy folds of the dog's eyelids, resulting in turned-in eyelids and eyelashes, irritating the eyeball. Almost permanent problems can be the various sorts of skin infections that develop as a consequence of the suboptimal hygiene of the deep wrinkles.

Below *When shar-pei dogs are selected against their "trademark" deep skin folds, the more streamlined look means they are less likely to suffer from skin infections.*

Malformation in Behavior ~

Digging

Marking

Chasing

Boredom

Barking

Right *Most of our dogs' misbehaviors are typical, species-specific behaviors that have strong biological roots.*

Selective breeding can have serious consequences, especially when it favors just a few traits or some extremity. Intense directional selection can decrease the genetic variability of the population to a dangerous level. Artificial selection in the case of dog breeds should balance narrowing the genetic variability in order to fix a desired trait and allowing for enough variability within a breed to ensure it a sound future.

MISBEHAVIOR—TYPICAL OR ABNORMAL BEHAVIOR?

Undesirable behavior or any behavior response that does not fit our lifestyle can be labeled as "abnormal" though it is not—not in a biological sense, which seems to be the only fair way to judge your pet's behavior. Often what we view as misbehavior is a typical natural behavior, a dog being a dog. Some can be truly annoying—for example, when, as part of his territorial behavior, a dog marks the wall during a visit to another apartment. Other behaviors that are also considered typical may be harmful (such as chasing bikes) and it is necessary to manage these. Digging in the flower bed or whining when bored can be considered as problem behaviors but these issues can

be solved by training. Behaviors that do lie outside the typical range and are considered pathologic may be the result of genetic predispositions, insufficient early socialization, or medical conditions.

EXCESSIVE BARKING

Barking is natural behavior in dogs, but constant barking can be irritating and it is easy to stigmatize such noisy dogs as neurotic. Such nuisance barking is one of the main reasons why people say they do not like dogs in their neighborhood. It is likely that dog domestication selected for enhanced vocalization, as in general dogs are noisier compared to their wild relatives. In addition, there were several breeding efforts to create ideal dogs for "treeing"—a method of hunting where dogs are used to force the prey up into trees, where they can be shot by hunters. Such dogs are selected for their habit not to cease barking at an animal even after it has escaped into a tree.

NOISE PHOBIA

The problem of extreme fear of loud noises has recently become common in dogs. They pant heavily, vocalize, and try to hide in dark and cozy spaces or blindly rush to escape, even hurting themselves in the process. Certainly, many normal living beings would search for a safe place in a heavy thunderstorm.

Some dogs are very sensitive by nature, experiencing general anxiety and being unable to cope with such noisy situations without help. Their genetic background could play a role in this. However, it may well be the fault of the owner, with the phobia developing as a kind of social learning when the puppy observes frightened people running into the house from a storm. Severe cases of fireworks fear or storm phobia may start with just a mild distress at extreme noises, then develop to a fear of any sudden noise, and finally increase to panic.

It is important to know that early intervention has the best chance of stopping the problem before it can get worse and spread to other situations.

Above *Some dogs with thunder phobia are also frightened of other loud noises, but others are only scared by storms. Many city dogs develop noise phobia after hearing fireworks.*

useful. For mother dogs eating up feces is not only normal but also important so that the litter can be kept clean. Considering that this is repulsive to human sensibilities and could potentially cause health problems, it is important that the owner discourages the young dog on the first occasion when such behavior is observed.

DOGS LEFT ALONE: SEPARATION ANXIETY

Many social species, including humans and dogs, exhibit distress responses when separated from attachment figures. This is normal and natural behavior for both infants and puppies.

Separation anxiety is relatively rare among outside-living dogs (though they are also devoted to their owners), which implies that there could be something more in the phenomenon than simply being overly dependent on the owner. Separation anxiety is much more common in dogs living in flats and apartments in towns and cities.

Companion dogs with separation anxiety typically eliminate, vocalize, or engage in destructive behavior when left alone in the apartment. Some breeds and some breeding lines are more prone to show separation anxiety than others. The role of inheritance is of crucial importance here, because only the mild, nongenetically based cases can be remedied easily by applying the training methods published on many experts' websites.

"Hyperattachment" to the owner has been assumed to be the main underlying cause of separation anxiety, but most results support a more complex

COPROPHAGIA

Maybe the most disgusting of all our dogs' bad habits is when they are happily searching for feces in the grass. Though there is a huge variability in this, dogs commonly feed on the waste of other dogs, cats, and other animals (including humans). Dogs eat poop for various reasons, primarily because they are scavengers, living off whatever they find explanation. For example, some dogs develop separation anxiety when they are 6–7 years old, while others seem to be born with it.

Since separation-related disorders in dogs seem to have considerably diverse underlying causes, the solution may need to include a specific combination of medication and behavioral therapy, which is adjusted for each individual case. Buying another dog might help reduce boredom, but usually does not solve the problem in the case of a true separation-related issue. As a short-term management, if possible, the use of a human dog sitter, or leaving the dog at a daycare center or a neighbor's, can reduce or solve the problem.

ABNORMAL AGGRESSION (RAGE SYNDROME)

Most dog owners have heard about it, spaniel fans fear it, and fortunately only very few dogs suffer from it—the condition commonly known as rage syndrome, more appropriately termed "idiopathic aggression." This disease has no known cause and is characterized by the unpredictability of the outbursts. It is not possible to predict when a companion dog may turn, without warning, into an aggressive beast and bite family members seriously. While most other types of aggression can be modified and reduced through expert training, rage syndrome is an extremely complex condition to deal with and sometimes euthanasia is the only solution. Genetic components are assumed to be responsible, as some breeds seem more prone to this condition, including cocker and springer spaniels, Bernese mountain dogs, and Lhasa apsos.

Right and below *Rage syndrome may occur in cocker spaniels, and can also develop, though rarely, in other breeds such as Lhasa apsos and Bernese mountain dogs.*

Myths & Misconceptions

Some myths and misconceptions have been around for centuries, while others have just been invented by today's "experts." The flood of tips can steer owners in wrong or sometimes dangerous directions. The most annoying of such urban legends convey false information claiming to be scientifically based.

Right *Dogs are highly social animals, they can learn socially (by watching you) and they can be rewarded socially (not just with food or a toy).*

OWNERS DON'T HAVE TO BE THE ALPHA

Fortunately, being a domesticated species, dogs were selected for dependence on a human partner, so they just need some basic, consistent, and kind training to spontaneously accept the leadership of the humans in the family. There are some breeds (including some livestock guard dogs and some lapdogs) and a few individuals from any breed that may show dominance aggression in conflict situations. However, these behavior problems need more complex treatment and the whole strategy of the owner should be changed. They need to implement a "Nothing in life is free!" program, where the dog has to earn everything he wants. Alpha-believers should also note that their belief in dominance aggression inevitably leads to contradictions within their argument: If our dogs' social behavior resembles that of wolves so strongly, then most family dogs would kill one another as soon as they are off the leash in dog parks.

ADULT DOGS CAN DEVELOP NEW BONDS

The ability to form attachment is usually associated with an early sensitive period while the dog is a puppy. Recent results have revealed that older service dogs and shelter dogs can establish proper attachment relationships with their new owners. The welfare aspects of these findings are of vital importance, because for a long time it was widely accepted that owners should obtain their puppies by the age of three months, otherwise no successful attachment bond could be developed. Although early socialization with humans can influence this ability, we now know that individualized bonds can develop throughout the life of a dog.

PUPPY TESTS ARE NOT NECESSARILY PREDICTIVE

Though owners and breeders accept the predictive value of puppy tests, there is no firm scientific evidence that single test items or the whole test battery have the potential to foretell the adult dog's personality. Observations carried out on a particular occasion may well be correct and applicable, but their lack of consistency and validation do not necessarily indicate the stable features of an individual. So, all that we know so far cautions us. Studies carried out on the predictive value of puppy tests of border collies, for example, reveal the minimal predictivity of such behavior tests.

THERE IS NO SUCH THING AS A SUPERDOG

Humans' desire for perfection is an understandable weakness. However, a dog cannot be both gentle and reliable with strangers and children and an effective sentry dog—as some beloved characters in movies are shown to be. The television superhero dogs are fictional: Dogs that are tough enough to do police work will not make trustworthy family pets, and vice versa. Even individuals from the most versatile breeds, such as the German shepherd dog, cannot simply be trained to be guide dogs for the blind or military dogs. Markedly different lines are bred within a breed for different purposes.

Below *Individuals within a breed do not necessarily have similar behavior traits. Selective breeding in German shepherds is practiced for dogs used either in the military or as guide dogs.*

The Scientific Study of Dogs ⟶

Dogs have become increasingly popular as subjects of behavioral research over the last two decades. This is because dogs can participate in behavioral experiments as easily as humans can. Researchers do not need to keep the animals in their laboratories; it is enough to recruit the owners and invite them to participate in an experiment with their dog.

But this is not the only reason for the growing body of research on dogs' behavior and cognition: Dogs and humans share the same natural environment, so researchers are able to observe the animals in their own environment very easily, an ideal condition for ethological studies.

EXPERIMENTS ARE CONTROLLED OBSERVATIONS

Ethologists aim to observe animals' behavior in their natural environment. However, to study their cognitive skills it is often necessary to test them in a more controlled laboratory setting. In this respect, the study of dogs is a remarkable case: Experiments are carried out in a dog scientist's laboratory that often looks like the living room of a human family, which is considered as the natural environment

for family dogs. The tests often involve some sort of playful interaction and can be considered a recreational activity for dogs and owners.

OBSERVING BEHAVIOR

Behavior is generally measured by direct observation. The main task of an ethologist is to provide a well-defined description of all actions performed by

Below *A dog is searching for food where the researcher points during an experiment. These and similar behavioral tests are entertaining activities for dogs and owners.*

the species. A hierarchically organized catalog of behavioral units is called an ethogram. Typically, human observers look for the frequency and duration of these behavior units.

At the moment, there is no generally accepted ethogram for the dog, although specific behavior units have been described as universally used. For example, looking at the owner's body could be considered as a behavior unit.

VIDEO ANALYSIS

The behavior of the dogs is usually video-recorded and analyzed using software specifically created for behavioral analysis. This analysis is performed by human experimenters and is prone to subjective interpretations. Moreover, recording behavioral variables relies on the perceptual abilities of the observer and can be influenced by his/her experience and emotional state.

To correct for these possible errors, two or more independent observers code the same videos and the agreement between them is calculated. Only if there is a high level of agreement between them can the analysis be considered reliable.

BLIND CODING

Human observers who are aware of the purpose of the experiment may have expectations about the outcome of the study and thus may be biased in how they judge the observed behavior. To avoid this kind of error, the coding is done by observers who do not know what the purpose of the study is, or are not aware of the treatment that was administered to the dog (blind coding).

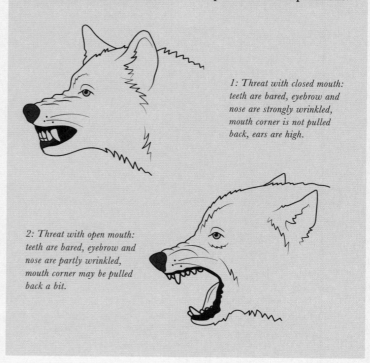

RELATED FACIAL EXPRESSIONS AS BEHAVIOR UNITS IN WOLVES

An ethogram is a hierarchically organized catalog of behavioral units. To create an ethogram it is fundamental to provide detailed descriptions of the actions performed by the species, such as these two examples of facial expressions.

1: Threat with closed mouth: teeth are bared, eyebrow and nose are strongly wrinkled, mouth corner is not pulled back, ears are high.

2: Threat with open mouth: teeth are bared, eyebrow and nose are partly wrinkled, mouth corner may be pulled back a bit.

AUTOMATIC MEASUREMENTS OF BEHAVIOR

New developments in technology have led to innovative methods for recording behavior. Small sensors can be applied on the body of the dog, for example, by placing them on a harness or on a collar, without interfering with the animal's behavior. These devices automatically record the direction of displacement, the speed, and the acceleration in three-dimensional space.

Above *At present, there is no commonly accepted ethogram for the dog, but ethologists rely on specific behavior units that have been described universally, such as the facial expressions in wolves.*

The Future for Dogs ~Q

Dogs have been around us for at least 16,000–32,000 years. For a large portion of humankind, this relationship has become stronger with time. However, human lifestyles have changed drastically with the shift to modern city-dwelling societies, and new cultural habits have emerged, facilitated by the rapid global exchange of information. At the same time, modern humans are becoming more and more distanced from nature—perhaps the dog is one of our last ties with the wilderness.

CHANGING VIEWS OF OWNERS

Although the love for dogs in many industrialized societies does not appear to be waning, there are changes in our views about this relationship. People are becoming practical, and life appears to have sped up and become much busier for many. Thus, potential dog owners prefer to invest less and less in their bond with dogs. When answering questions in a survey on the "ideal pet," the majority

Below For some owners the dog's happiness is of foremost importance. Accordingly, these dogs are regarded as family members, and can do mostly what they want.

of Australian dog owners indicated that they preferred a dog that is desexed, safe with children, fully housebroken, friendly, obedient, and healthy. While these traits are obviously important, they do not come without a lot of social investment in the socialization and training of the dog companion.

HUMANIZATION OF DOGS

One modern trend is to regard dogs as "little furry babies." Dogs are dressed up, carried around in cars or even baby carriages, and taken to wellness centers. Although research has established many functional parallels between companion dogs and human babies, this does not mean that they should be treated as equals. Even if one could justify the treatment of dogs as babies, one should never forget that the dog is an animal species in its own right, not an accessory to human self-realization.

LICENSE FOR DOG OWNERSHIP

While many dog owners consider their dog more as a baby, the mistreatment of many companion dogs has led others to argue that owning a dog should be subject to a license. Indeed, this idea has become law in Switzerland, where prospective dog owners must participate in both theoretical and practical seminars at accredited dog schools before they are allowed to get a dog.

This new trend may indeed increase the welfare of dogs because most dog owners from the last two to three generations have really lost contact with nature, so they must be taught what an "animal" is, and the ways in which one should deal with it.

Above *Dogs should not be treated as little humans. Although in most cases dressing up a dog causes no harm, some dogs may feel bad if they are laughed at.*

Below *Many dog owners are so distracted by present-day technology that they ignore their four-legged companion.*

MISUSE OF
ETHICAL ARGUMENTS

People not trained in biology often do not understand the limitation of arguments based on analogies. Many people are not happy that they have to work to earn money. So many think that making dogs work is not appropriate. One needs to consider that what may be perceived as work in the case of humans is rather the typical activity that an animal has evolved to do. Animals and dogs are programed to receive stimulation from the environment and to respond to it. Understimulated dogs or individuals that are not allowed to perform according to their natural behavior may develop behavioral and mental malformations.

In addition, the general impression from studying dogs is that a well-trained dog that has been chosen for a particular task enjoys interacting with its owner. They would probably suffer if they could not interact with humans. Working, for them, is closer to some kind of social engagement than a form of hard labor. It is also important, however, that they receive appropriate positive social feedback for being such a devoted companion, and have enough time to rest and refuel.

IS SPORT FOR THE
DOG OR THE OWNER?

Some dog owners like to engage their dog in sports, and these, in some cases, do indeed challenge the health and welfare of the dogs. These include greyhound and sled-dog racing, or weight pulling. Even if these activities may once have been considered typical for some breeds of dog, the competition and practice for such races may expose dogs to unnecessary discomfort, and may compromise their health, possibly resulting in early death. Unfortunately, little research into this area has so far been carried out. But a negative outcome is very likely. Society needs to

Below *Dog racing, of sled dogs, greyhounds, and other breeds, has been around for centuries. However, it is doubtful whether fast running over many miles is really a good sport for dogs.*

consider to what extent it allows such competitions. For example, greyhound racing is currently illegal in many countries and most US states.

The other side of the coin is that competitions in general may provide one way of keeping breeds and dogs in general in good genetic condition because breeding can be based on the most successful individuals. However, such decisions should be based on problem-solving skills and interactive behavior rather than extreme physical performance.

COMPETITION FROM TECHNOLOGY

While dogs enjoy a specific status as our companions, the need in humans to have a social partner, such as the dog, can now be satisfied in many different ways. In the early 2000s the Japanese media company Sony marketed a small toy robot called AIBO (the name meaning companion/pal in Japanese). Although

Below AIBO, *the robot dog, was produced as a toy for children. It performed many different dog-like actions but it moved very slowly and had limited sensory abilities.*

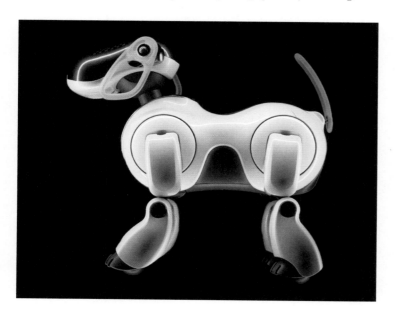

AIBO was able to display a wide range of doglike behaviors and could also learn to perform simple actions on command, it was never taken seriously as a competitor to real dogs, and disappeared from the market a few years later.

There is, however, a more serious contender in the race to become human's new chosen companion—the cell phone. People spend more and more time on their phones, and they use them habitually when walking their dogs. As technology advances, it may also be that other more agile alternatives will replace dogs in the long term if people cease to strive for a real social relationship with a devoted companion.

TECHNOLOGY MAY ALSO HELP

Recent developments in wearable sensors offer novel possibilities to monitor the behavior and health status of the dog, and to assist vets in making better diagnoses, tracing changes in the dog's behavior, or even alerting the owner in some cases. For example, such sensors, specially designed for dogs, can recognize scratching behavior and can notify the owner if the dog is showing an allergic reaction while at home alone. Similarly, sensors can detect an unexpected epileptic seizure and report it to the owner. By using this technology, the owner can be aware of the dog's condition if he/she is not at home. Importantly, technology should never replace the owner; rather it should act as a catalyst for strengthening the dog–owner relationship.

CHAPTER 6

A Directory of Dog Breeds

How Contemporary Breeds Exist ∽ℓ

Left and below *Few people can resist a puppy, whether it is a mixed breed (left) or pure bred (French bulldog, below).*

All dog owners love dogs but many of them love some dogs much more— purebred dogs. In several branches of the dog-loving society it is noticeable that mixed-breed dogs (also called mongrels) are regarded as outcasts. There are also many devoted dog enthusiasts whose hearts go out to those outcasts and who rescue them whenever they can. The care for these individuals is, however, equally important because the diverse gene pool present in today's mixed breeds may provide a reservoir if inbreeding becomes a very serious issue.

KENNEL CLUBS

Kennel clubs and local breeding organizations are the guardians of the fate of "pure" dog breeds. In many countries, large alliances take the responsibility for providing guidelines and rules for breeding "purebred" dogs and this often results in marked inbreeding. It is unfortunate that many of these organizations grew traditionally from the campaigns to establish competitive dog shows because even today most of them are more concerned about the look of the dogs rather than their physical and mental health.

Some kennel clubs, however, do a good job if they recognize that they need to intervene in order to maintain dogs' welfare in general, regardless of whether dogs are purebred or mixed. Kennel clubs can be a huge assistance to dogs if they support the development of DNA tests designed for detecting specific genetic diseases, endorse physical and behavioral screening for dogs to become breeders, and encourage selective breeding and precise documentation of the pedigree.

Left *English bulldogs were subjected to extreme selection that made their head shape problematic for breathing and eating. Now some kennel clubs aim to reverse this process by hybridization.*

BREED-SPECIFIC LEGISLATIONS

Dog keeping has been regulated by different laws over time and from country to country. Most famous and disputed are breed bans, the prohibition of keeping and/or breeding certain breeds. In most cases, these regulations were imposed after some serious incidents when a dog killed an adult or child, and authorities argued that such bans would decrease dog biting. In most countries, the so-called "fighting dogs" (such as the pit bull terrier) and large mastiff-like breeds (such as the dogo Argentino) are banned. Although individuals of such breeds are likely to have participated in some dog-biting incidents, these breeds are relatively rarely involved. The majority of dog biting is caused by the much more common dog breeds. Thus, breed bans do not generally lead to a decrease in the number of dog bites.

RESPONSIBLE DOG OWNERSHIP

The so-called "responsible dog ownership" movement has convinced many people of the advantages of dog neutering. The main argument was that there are already too many dogs (especially free-ranging dogs), and in this way the owner can prevent the birth of unwanted puppies. It was also indicated that neutering may protect from specific diseases.

However, it is equally important that there are enough breeding individuals in the reproductive population producing the next generation of puppies. If the breeding population is too small, then increased inbreeding could lead to general deterioration of the breed and the emergence of specific illnesses. Thus, dog breeding organizations, dog breeders, and dog owners should develop a common strategy that ensures a genetically diverse population of healthy dogs.

Above *The strength and hunting and guarding traits of the dogo Argentino make it an unsuitable breed for keeping in small apartments.*

Modern Dog Breeding

Modern dog breeding started with the establishment of kennel clubs during the second half of the nineteenth century. The generic groups based on main functions, such as hunting dogs, herding dogs, or guard dogs, were divided into many subgroups based on subtle differences in their roles and behavior.

HUNTING DOGS

The major categories of hunting breeds include gundogs, terriers (earth dogs), and hounds, but even these groups can be broken down further. Gundogs can be classified as *retrievers*, whose role is to visually spot and remember the location of downed birds and return them to the hunter; *flushing spaniels*, which, in contrast to pointers, are trained to find and flush out the game, mostly birds; and *pointing breeds*, such as setters and pointers, which use their noses to find the game and then stalk up on it, pointing at the hiding bird.

Terriers and hounds were selected to hunt on their own without the direct guidance of the owner. *Terriers* were bred to kill small predators or vermin, mainly rats, rabbits, and foxes both over and under the ground, but typically grabbing the prey in tunnels or burrows where people could not reach them. Hounds were classified depending upon the primary sense they used to locate quarry and fall into two types: *sight hounds* (for example, greyhounds) detect and pursue the prey by sight and kill it, and *scent hounds* (such as beagles) follow the scent of tracks and are used to trail the game.

Below *Gun dogs such as the Hungarian vizsla make ideal companions if their owners can devote the time and energy they need.*

SHEPHERD DOGS

Shepherd (or pastoral) dogs include working breeds developed to help the shepherd herd and/or protect livestock. *Herding dogs* are small to medium-sized breeds, active and attentive, which work in close contact with the shepherd. *Sheepdogs* (border collie, kelpie, puli) have been bred for herding sheep, while *cattle dogs* or *heelers* (Australian cattle dog) were mainly developed for use as herders and drovers of cattle.

Different types of herding dog apply different strategies and behaviors to control the livestock; heelers and some Continental sheepdogs drive the livestock from behind, moving them away from the herder. Border collies position themselves along the sides or at the front of the flock, controlling them by eyeing (staring), which together with their typical crouching posture resembles the characteristic stalking behavior of predators.

To protect livestock from predators and thieves, large, powerful, and self-reliant dogs were bred and have been used for centuries in Europe (the Great Pyrenees, for example) and Asia (the Akbash and Caucasian shepherd dog, which protect against bears, wolves, and jackals as well as human intruders.) They tolerate other dogs and people they know. Some, such as the kuvasz, can function as both a livestock guard and a home security guard.

These giant breeds behave very gently with lambs and goat kids. They seem to bond to their stock (rather than to the shepherd) and sometimes exhibit nurturing behavior toward young animals of the other species. Most importantly, they provide predator-friendly control by giving a graduated response—the dogs begin with barking to try to frighten the predator, followed by posturing or charging, and finally attack only if necessary. Livestock guard dogs provide long-term protection as predators do not become habituated toward them.

Above *Large livestock guard dogs, such as the Caucasian shepherd dog, have not been bred to be companions; their natural personality makes them distrust strangers and exhibit territorial behavior.*

Mixed-breed Dogs (Mongrels) & Crossbreeds ～♪

There is no universally accepted categorization of dogs that do not belong to a specific breed. The "mixed breed" label usually means dogs that do not belong to any recognized "pure" breed and whose ancestry is complex or unknown. The term "mixed breed" is often preferred as a label because it has no negative connotation, unlike "mutt" or "mongrel." However, the latter are used mostly in reference to dogs that have a totally uncertain ancestry going back many generations. In contrast, crossbred dogs are intentionally bred by humans as hybrids of recognized breeds.

DESIGNER DOGS

These canines are crossbred with a purpose, either to optimize the best qualities of parents from two different breeds or to fulfill the needs of people who are impressed by a pet that seems to offer them an elevated social status, such as other "designer" goods do.

Labradoodles were among the first designer dogs, bred from Labradors and poodles to be service dogs with a low-shed coat. Puggles are bred from beagles and pugs—usually the offspring has the muzzle of a beagle, which can eliminate breathing problems. However, it is also

Above and right *Designer dogs, like labradoodles (top), black Russian terriers (opposite top), puggles (above), and lurchers (right) are crossbred with a purpose, usually to optimize the best qualities of the parents or to develop a new type of dog.*

possible for a puggle to inherit a short snout and a hunting instinct from, giving it a respiratory system that is not able to handle all the exercise the dog needs.

Although the history of designer dogs began in the late twentieth century, deliberate crosses had important roles in certain parts of the world much earlier. A lurcher is the offspring of a sight hound

and most commonly a pastoral breed or a terrier. The story goes that in the Middle Ages the English and Scottish governments banned commoners from owning sight hounds, and lurchers were bred to avoid legal complications. However, it is much more likely that the aim of the cross was to produce a fast sight hound with increased trainability skills, suitable for poaching rabbits, hares, and game birds.

Designer dogs (crossbreeds) are not recognized by traditional breed registries. But if they are bred together for a certain period of time, and their breeding is well documented, they may eventually be considered a breed of dog if the dogs show a uniform look.

For example, in the development of the black Russian terrier during the 1940s and 1950s, several breeds were used, including the Airedale terrier, the giant schnauzer, and the Rottweiler, and it was recognized by the Fédération Cynologique Internationale in 1983 and the American Kennel Club in 2004.

WHAT CAN A DNA TEST TELL ABOUT A MIXED-BREED DOG?

DNA tests vary between providers and need to be validated against different dog populations. DNA tests examine an individual's genetic markers against a database of markers for several hundred dog breeds. A computer program calculates a probable "pedigree." As databases are expanded, these tests become more reliable but, because of the close relationship between breeds, many chance effects are possible. Thus, any results should be considered as a possibility and not as an absolute truth because not all mixed breeds are descendants of purebred dogs.

Left *The shepsky is a crossbreed between the German shepherd dog and the Siberian husky.*

THE PROPORTION OF MIXED-BREED TO PUREBRED DOGS ACROSS THE WORLD

In scientific databases based on convenient samples, mixed breeds (dogs for which the pedigree is not known and which do not belong to any recognized breeds) compose approximately one-third of the dogs living in human families. However, these proportions are not representative, and suggest that owners of purebred dogs more frequently volunteer to participate in behavior tests or fill in questionnaires. According to more precise estimates, the relative proportion of such dogs varies among the more industrialized countries from approximately 30 percent in the UK and Germany to 50 percent in the United States and Australia.

In other cultures, where owning a dog in the way people in industrialized countries do is rare, the majority of dogs are free-ranging with unknown origin. Consequently, such mixed breeds represent the majority of dogs worldwide, although most of them live as free-ranging stray dogs or pariah dogs.

MIXED BREEDS IN THE FAMILY

A recent study detected a number of other systematic differences between mixed-breed and purebred companion dogs in a sample of 6,000. The mixed-breed dogs were found to be older in the sample, there were more females among them than among the purebreds, they were more likely to be neutered, owners acquired them at an older age, they received less training, they were more likely to be kept only indoors, and more

were kept as single dogs than in the case of purebred dogs. More women kept mixed breeds than men, owners were younger, had a lower level of education, and had less previous experience with dogs than the owners of purebreds.

DIFFERENCES BETWEEN CROSSBREEDS, MIXED BREEDS & PUREBREDS

Data suggest that, on average, mixed breeds and crossbreeds have some health advantages over purebreds, probably because they are less inbred. As they have much higher genetic variation, they are less likely to suffer from inherited genetic diseases and more likely to live longer than purebreds (owing to hybrid vigor—see opposite).

Behavioral studies detected differences between mixed breeds and purebred dogs. Mixed-breed dogs were generally reported to be more disobedient, more nervous, more excitable, and more fearful. Excessive barking was also more frequent in their case.

They were also reported to be more aggressive toward unfamiliar people, more sensitive to touch, and had an increased risk of developing behavior problems (such as noise phobia) than purebreds. One may assume that most differences emerge because mixed-breed

Above and right *It is difficult to predict the adult appearance and personality of a crossbred puppy, such as the pomsky (above) and cockapoo (right). The dog may have characteristics of both parents or of either of them.*

Left *A bosky, a Siberian husky and boxer mix, requires lots of exercise and training.*

Far left *The Czechoslovakian wolf-dog is a hybrid of the German shepherd dog and the Carpathian wolf.*

dogs experience less optimal early socialization and their genetic constitution also differs.

It is possible that a large proportion of present-day mixed breeds with unknown genetic histories originate from populations that have been under continuous selection for independent survival skills, making them more independent, assertive and more nervous/alert. In contrast, dog breeders generally selectively breed dogs that make good human companions, focusing on favorable (calmer) behavior characteristics. Despite the different selective forces, numerous mixed-breed individuals obviously do make ideal companion dogs, as the magnitude of the differences between mixed breeds and purebreds are small.

HYBRID VIGOR

Hybrid vigor is also known as heterosis or enhancement of outbreeding. The first generation of crossing dogs from two different breeds (or lines within a breed) produces hybrids with usually positive overall effects on health and biological functions. The explanation is that these individuals more likely inherit different gene variants (heterozygosity) that make them more resistant to environmental challenges, including pathogens.

Heterozygosity may be advantageous in other cases as well. Whippets with a single copy of a mutated gene that affects muscle composition are more muscular, and are among the fastest dogs in racing. However, whippets with two copies of the same mutation (homozygosity) develop too much muscle, which makes them rather slower.

French bulldog

HEIGHT
10–14 in./24–35 cm

ORIGINAL FUNCTION
companion

EXERCISE moderate;
several short walks

HEALTH ISSUES many;
proneness to heat
exhaustion, hip dysplasia,
respiratory distress,
allergies, patellar luxation

AFFECTION high

TRAINABILITY moderate

CARE NEEDS moderate;
care of skin (wrinkles)

WITH PETS good

WITH CHILDREN
good/excellent

PROTECTIVITY low

FOR NOVICE OWNERS good

Profile Despite his name, the French bulldog originated in England in the nineteenth century, when he was bred to be a toy version of the bulldog. His unique appeal—with large bats ears, short nose, and expressive eyes—has made the "Frenchie" one of the most popular pets in Europe and the United States. This little bulldog is small but compact and substantial in build with a powerful, muscular body. His short coat can come in brindle and fawn colors with or without white spotting. Unfortunately, the extreme snub nose and the unnatural proportions of the body result in several health issues.

Behavior and upkeep French bulldogs are loving companions, can be surprisingly smart, but also have the tendency to be stubborn. In contrast with their sullen facial expression, Frenchies have a gentle nature and can get along well with everyone, including children. Some can be possessive but they are rarely aggressive. While young they are rather playful, later they tend to be lazy, but still need regular exercise to keep them fit and prevent weight problems. French bulldogs snore loudly.

COUNTRY OF ORIGIN France

Jack Russell terrier

HEIGHT
10–12 in./25–30 cm

ORIGINAL FUNCTION
earth dog

EXERCISE much;
long walks, sport, tasks

HEALTH ISSUES few;
eye disorders

AFFECTION moderate

TRAINABILITY high

CARE NEEDS low;
occasional hand-stripping

WITH PETS fair

WITH CHILDREN good

PROTECTIVITY low/
moderate

FOR NOVICE OWNERS
not suggested

Profile The Jack Russell terrier was named after its originator, the reverend John (Jack) Russell, an English parson in the early 1800s. These small terriers were bred to hunt vermin. Being selected for temperament and ability rather than consistency in type, there was a considerable variation in size, shape, and type. They come in three coat types, all of which are virtually weatherproof: smooth, broken, and rough. Their color is solid white, or predominantly white with black, tan, or brown markings.

Behavior and upkeep The modern Jack Russell is a small, agile, active dog who needs to use his mind as much as his body. He is brainy, tireless, and a trainable partner for work or sport, but not the pet for everyone. He can handle large, loud families and adapts well to apartment living if the owner is willing to devote the time and energy to training. Jack Russells can be fierce with unfamiliar dogs without proper socialization. If exercise is not provided on a daily basis, this curious and highly energetic dog may bark excessively or become destructive out of frustration.

COUNTRY OF ORIGIN England

Dachshund

HEIGHT (standard)
8–9 in./20–23 cm

ORIGINAL FUNCTION
earth dog

EXERCISE moderate;
several short walks

HEALTH ISSUES some;
paralysis, keratitis,
ear infections

AFFECTION high

TRAINABILITY moderate

CARE NEEDS moderate;
regular ear cleaning,
trimming of nails if needed

WITH PETS fair/good

WITH CHILDREN fair/good

PROTECTIVITY low/
moderate

FOR NOVICE OWNERS good

Profile This short-legged, long-bodied, hound-type breed was originally created to hunt badgers and other tunneling mammals (the German *Dachs* means "badger"). The dachshund is bred in three sizes: standard, miniature, and rabbit (kaninchen), based on a chest measurement and weight. (In the United States, they are either miniature or standard, based on their weight.) They come in three coat varieties: smooth/short-haired (the original type), wirehaired (created through crosses with terriers), and long-haired (created through crosses with spaniels).

Behavior and upkeep Dachshunds are smart dogs with an independent nature and playful spirit. Housebreaking is sometimes not easy, but patience and consistency can help. They like to bark and often have a surprisingly deep voice that makes them sound larger than they are. They tend to bond closely with a single person and may even become jealous of their owner's attention.

COUNTRY OF ORIGIN Germany

Miniature schnauzer

HEIGHT
12–14 in./30–35 cm

ORIGINAL FUNCTION ratter

EXERCISE much; long walks, sport, tasks

HEALTH ISSUES some; allergies, bladder stones, epilepsy, diabetes, pancreatitis

AFFECTION high

TRAINABILITY
moderate/high

CARE NEEDS low/moderate; trimming/hand-stripping 2–3 times a year

WITH PETS fair/good

WITH CHILDREN fair

PROTECTIVITY moderate

FOR NOVICE OWNERS fair

Profile This small, terrier-like breed has been used as a ratter on German farms for many centuries. He descends from the much larger standard schnauzer, and several small breeds contributed to the final, tiny variant. Their wiry double coat comes in solid black, salt-and-pepper (gray), black and silver, or solid white. According to European legislation, the ears and tail cannot be cropped. They do not shed if stripped regularly.

Behavior and upkeep Miniature schnauzers are intelligent, highly trainable, and are among the best in agility contests. However, due to their independent nature, they can also be stubborn, so owners must show them that they really mean what they say. They are alert watch dogs but not fighters. Proper early socialization and enrolling in sport activities can ensure that the puppy grows up to be a well-rounded dog. If they are not having enough exercise, they tend to chase cats, bark excessively, or dig holes in the garden.

COUNTRY OF ORIGIN Germany

Chihuahua

HEIGHT 6–9 in./15–23 cm

ORIGINAL FUNCTION unknown

EXERCISE moderate; several short walks

HEALTH ISSUES some; patellar luxation, alopecia (bald patch), prone to dental problems

AFFECTION high

TRAINABILITY moderate/high

CARE NEEDS low/moderate; weekly brushing, trimming nails

WITH PETS fair/good

WITH CHILDREN fair

PROTECTIVITY moderate

FOR NOVICE OWNERS fair

COUNTRY OF ORIGIN Mexico

Profile The Chihuahua is said to descend from ancient Mexican breeds, but its exact origin remains unknown. Being the world's tiniest breed, the dog is especially well suited to urban life as a "pocket pet." The breed's most distinctive features are the apple-shaped head with pointed muzzle, large, expressive eyes, and upright, batlike ears. Both smooth and long coats are common, and they can come in any color but are typically fawn, chestnut, black and tan, blue, or tricolor. Chihuahuas love to be warm, but they shiver not only when they are cold, but also—as with many other tiny breeds—when they are excited or nervous.

Behavior and upkeep Ideally, they are alert, bold, highly intelligent little dogs, but many are shy, feisty, and snappy toward strangers or children. Moreover, they can be dangers to themselves, often posturing toward bigger dogs, or even biting out of fear. Excessive barking is a typical behavioral problem. They need a patient and consistent hand to train them.

Yorkshire terrier

HEIGHT 6–9 in./15–23 cm

ORIGINAL FUNCTION ratter

EXERCISE moderate; several short walks, tasks

HEALTH ISSUES some; keratitis, patellar luxation, eye diseases, dental problems

AFFECTION high

TRAINABILITY moderate/high

CARE NEEDS moderate

WITH PETS good

WITH CHILDREN good

PROTECTIVITY moderate

FOR NOVICE OWNERS good

COUNTRY OF ORIGIN Great Britain

Profile The Yorkshire terrier was originally developed in England and used by miners as an efficient ratter. Later they were discovered by dog fans to be charming companions and have been bred down in size to make a pretty and portable pet. The present-day Yorkie, being just slightly larger than his former enemy, would not have much chance as a terrier, even though some of them have retained their courage and spirit. Due to their career as a companion, they have developed beautiful long, straight, silky hair that comes in metallic blue and rich golden tan. Yorkies are much more delicate than other terriers; they can be seriously injured even in peaceful play with giant breeds.

Behavior and upkeep The ideal Yorkie is a jolly and playful companion that is highly devoted to his owners. He is alert and responsive, affectionate with the family, and often standoffish with strangers. While highly trainable, Yorkies should be socialized and trained young in life. Without extensive socialization or due to genetic predispositions, they can be scrappy, nervy, or yappy.

Shih tzu

HEIGHT < 11 in./27 cm

ORIGINAL FUNCTION companion

EXERCISE moderate; long walks

HEALTH ISSUES some; allergies, eye problems, heat stroke

AFFECTION high

TRAINABILITY moderate

CARE NEEDS moderate/high; regular brushing, care of eyes, ears, and wrinkles

WITH PETS excellent

WITH CHILDREN excellent

PROTECTIVITY low

FOR NOVICE OWNERS excellent

Profile The shih tzu descends from the "little lion" dogs in China and Tibet and is a close relative of the Lhasa apso. He has some similarities with other small Oriental breeds, but his distinctive feature is his chrysanthemum-like face. The long, silky, abundant coat comes in many colors, mostly solid black, black and white, gray and white, or red and white. A white tip on the heavily plumed and gaily carried tail and a white blaze on the forehead are highly desirable. Due to its short nose, the shih tzu tends to wheeze and snore.

Behavior and upkeep Only their looks resemble those of a little lion—they have a charming and gentle character. They do very well in small apartments, and their docile personality makes them a good companion even for first-time owners or children (but not for toddlers). As lapdogs, shih tzus are happy with regular gentle walks and do not need much activity. Some can be alert and make good watchdogs; they do not bite, rather take a reserved, wait-and-see approach to strangers.

COUNTRY OF ORIGIN Tibet

Pug

HEIGHT
10–14 in./25–35 cm

ORIGINAL FUNCTION
companion

EXERCISE low;
several short walks

HEALTH ISSUES many;
patellar luxation,
encephalitis, epilepsy, nerve
degeneration, allergies,
prone to eye injuries

AFFECTION high

TRAINABILITY moderate

CARE NEEDS moderate;
wiping eyes and wrinkles
regularly

WITH PETS excellent

WITH CHILDREN excellent

PROTECTIVITY low

FOR NOVICE OWNERS
excellent

COUNTRY OF ORIGIN China

Profile Originating in China and being the lapdogs of emperors, the pug has a long
and adventurous history. This sturdy, compact dog is known for its distinctive features
of deep wrinkles around big, dark eyes and a flat, round face. His coat comes in silver,
apricot, fawn, or black. Pugs have a characteristic undershot jaw and, due to this
specific feature, do not tolerate extreme weather. Unfortunately, pugs are not among
the healthiest breeds, so finding a reputable breeder is key to avoiding genetic
diseases.

Behavior and upkeep These little clowns do not need a lot of exercise but do
need daily walks to keep them healthy. Pugs are typical couch potatoes and greedy
eaters, so they are prone to become obese. They are playful, affectionate, and
mischievous, but they can also be lazy and stubborn, which makes training challenging.
Their quiet and easygoing nature makes them ideal for inexperienced owners.

Staffordshire bull terrier

HEIGHT
14–16 in./35.5–40.5 cm

ORIGINAL FUNCTION
fight dog/terrier

EXERCISE moderate/much;
long walks, tasks

HEALTH ISSUES some;
mast cell tumors, cataracts,
hip dysplasia, epilepsy,
heat exhaustion

AFFECTION high

TRAINABILITY
moderate/high

CARE NEEDS low

WITH PETS good

WITH CHILDREN
good/excellent

PROTECTIVITY
low/moderate

FOR NOVICE OWNERS good

Profile Descended from dogs bred for bull baiting and later pit fighting, the Staffordshire bull terrier is a close relative of the American Staffordshire terrier. This small-built but muscular breed is well balanced, showing great strength for his size. His general appearance radiates power and gives the impression of determination and action. Despite his athletic body, he can quickly become obese if food intake isn't monitored carefully. Coats come in red, fawn, white, black, blue, bridle, or any one of these colors with white.

Behavior and upkeep "Staffbulls" are self-assured and good-natured as well as attentive and willing to please. Those who do not know him would be surprised that this high-energy breed has a real soft side. Intelligent, affectionate, and reliable, the Staffbull is famous for combining toughness with playfulness. Their boisterous personality requires a firm, consistent, and kind handler. As long as the dog comes from a reputable breeder with a gentle bloodline, and is socialized properly, he will also be reliable with other dogs and pets.

COUNTRY OF ORIGIN Great Britain

Cavalier King Charles spaniel

HEIGHT
12–13 in./30–33 cm

ORIGINAL FUNCTION
companion

EXERCISE moderate;
several short walks

HEALTH ISSUES some;
upper airway syndrome
patellar luxation, entropion,
retinal dysplasia

AFFECTION moderate/high

TRAINABILITY moderate

CARE NEEDS moderate;
weekly brushing, checking
ears for infection, trimming
nails

WITH PETS excellent

WITH CHILDREN excellent

PROTECTIVITY low

FOR NOVICE OWNERS
excellent

Profile In contrast to his name, this small breed has never hunted as a spaniel, but was a luxury lapdog bred purely for companionship, especially for the children of royalty. Cavaliers have long, feathered ears, silky, often wavy coats, and come in shades of solid ruby, black, and tan as well as tricolor and Blenheim. Aside from being one of the most popular breeds, he has all the most common loveable features that people are looking for in dogs—loyalty, expressive eyes, childlike face, attention, desire to please, and an easygoing nature.

Behavior and upkeep Though they can live happily in an apartment, Cavaliers want to be included in all activities, both outdoors and indoors. With his begging eyes and gentle expression, a Cavalier can easily manipulate the heart of most owners, which is problematic because they gain weight easily. For their training, a gentle hand is required, as harsh treatment would break their sensitive personality. Separation anxiety is the most common problem with this breed, so they should not be left alone for long periods of time.

COUNTRY OF ORIGIN Great Britain

Bulldog

HEIGHT
12–15 in./30–38 cm

ORIGINAL FUNCTION Bull
baiting/fight dog

EXERCISE low; several
short walks

HEALTH ISSUES many;
respiratory distress, eye
problems, deafness, cancer,
hip dysplasia, heat stroke

AFFECTION high

TRAINABILITY
low/moderate

CARE NEEDS much; care
of eyes and wrinkles

WITH PETS excellent

WITH CHILDREN excellent

PROTECTIVITY low

FOR NOVICE OWNERS
excellent

Profile This molossoid breed was originally used for bull baiting, then fought with other dogs in the pits to evolve finally into a funny-shaped pet with a distinctive expression on his wrinkled, childish face. He is rather low in stature, but disproportionally broad, powerful, and compact with a relatively large, round head. His coat comes in many colors, brindle shades of red with or without a black mask, white, and pied. Considering the many potential health issues related to the breed, it is an absolute necessity to look for a reputable breeder who avoids extremities and tests the breeding stock for genetic diseases.

Behavior and upkeep What remained from the one-time ferocious fighter, is a loyal, affectionate, dependable companion. Some may be alert guard dogs but they are never vicious or aggressive. Though fierce in appearance and still courageous and confident in most situations, the modern bulldog is gentle and affectionate.

COUNTRY OF ORIGIN Great Britain

West Highland white terrier

HEIGHT ~11 in./28 cm

ORIGINAL FUNCTION terrier

EXERCISE moderate; long walks

HEALTH ISSUES some; anemia, skin disorders, hip dysplasia

AFFECTION moderate

TRAINABILITY moderate

CARE NEEDS moderate; regular stripping and/or clipping of hair

WITH PETS good

WITH CHILDREN good

PROTECTIVITY low/moderate

FOR NOVICE OWNERS good

Profile The "Westie," as it is nicknamed, originates from Scotland and was bred for hunting rats, foxes, and other vermin. Once, he was simply the white version of the cairn terrier. The Westie inherited his relatively docile personality from the cairn, but, because of his distinctive color, gained greater popularity and gradually embarked on the path to becoming a lapdog. He is still a sturdy dog, a balanced combination of strength and activity. The outer coat is harsh and the undercoat is short and soft.

Behavior and upkeep The Westie is one of the most amicable and tractable members of the terrier group. This lively and strong-willed breed needs a significant amount of exercise and mental stimulation, especially during the first year. Westies are loving family companions, but can be stubborn and refuse to listen at times. They are independent and can be insolent and possessive, and are prone to bark a lot.

COUNTRY OF ORIGIN Great Britain

Whippet

HEIGHT
17–20 in./44–51 cm

ORIGINAL FUNCTION racing

EXERCISE much/moderate; long walks, sport

HEALTH ISSUES some; eye and skin disorders, get cold easily, thin skin is vulnerable to injuries

AFFECTION high

TRAINABILITY moderate

CARE NEEDS low; trimming nails if needed

WITH PETS fair

WITH CHILDREN fair/good

PROTECTIVITY low

FOR NOVICE OWNERS good

Profile Called the poor man's greyhound, the whippet is a medium-sized, lean, muscular, and graceful sight hound that resembles a smaller version of the greyhound. Whippets were bred for speed and prey drive, with the aim of developing a dog that would be extremely efficient in chasing and killing rabbits—mainly not in the field but in contests organized by working men. Whippets are sprinters; at short distances some could outrun a greyhound.

Behavior and upkeep Though they are extremely energetic outside and need regular runs or play with other dogs at top speed, whippets make surprisingly quiet, adaptable, and calm companion dogs at home. As with all greyhounds, they are prone to chase and have a strong prey drive, so they can get along well with small pets only if they are reared together. Off-leash exercise is especially important when they are young. They are affectionate and gentle with their owners but can be shy with strangers, so they need extensive socialization.

COUNTRY OF ORIGIN Great Britain

Border terrier

HEIGHT
10–11 in./25–28 cm

ORIGINAL FUNCTION
earth dog

EXERCISE much;
long walks, sport, tasks

HEALTH ISSUES few

AFFECTION moderate

TRAINABILITY
moderate/high

CARE NEEDS low; (hand)
stripping twice a year

WITH PETS good

WITH CHILDREN excellent

PROTECTIVITY low

FOR NOVICE OWNERS
excellent

COUNTRY OF ORIGIN Great Britain

Profile Originating from the border of England and Scotland, this plucky working terrier was bred to kill foxes. His body is strong but athletic, combining activity and endurance with strength. The border is known for its distinctive otter head, and a sturdy tail that provided hunters with a "handle" to pull them out of the burrow. His harsh and dense double coat is naturally dirt- and water-resistant. Its color comes in wheaten, red, grizzle/blue, and tan.

Behavior and upkeep The popularity of this racy breed is due to their complex personality; even though in the field they are as tough as any other terrier, at home they are affectionate, obedient, and eager to learn. They are a good match for an active owner, especially one who is happy to get involved in dog sports, such as agility. Due to their inquisitive, busy, friendly, and biddable nature, they can easily handle even a noisy, chaotic household.

Beagle

HEIGHT
13–16 in./33–40 cm

ORIGINAL FUNCTION
scent hound

EXERCISE moderate/much;
long walks

HEALTH ISSUES some;
epilepsy, allergies,
hip dysplasia

AFFECTION moderate

TRAINABILITY moderate

CARE NEEDS low/moderate;
ears are prone to infections
so need regular cleaning

WITH PETS excellent

WITH CHILDREN excellent

PROTECTIVITY low

FOR NOVICE OWNERS good

Profile The smallest British pack hound was bred to hunt hare or rabbit, while followed by hunters on foot. Though foxhounds are famous for their independent spirit and are more eager to follow their noses than their owners, the beagle is one of the most popular breeds worldwide. The short, dense coat comes in all hound coloring—tricolor or red and white. Most beagles have a white-tipped tail.

Behavior and upkeep The sturdy, bustling beagle requires plenty of activity. They are bold and alert, showing no aggression or timidity. As pack dogs they are exceptionally tolerant of other dogs, but also very sensitive to separation. In spite of their easygoing nature with children, they are not necessarily the best pet for a child because their training is not without challenges. All hounds were selected for following a scent on their own, so it is no surprise that they rarely respond to calls without massive training. They are full of energy while young but they can be lazy and incurious later, especially without regular long walks.

COUNTRY OF ORIGIN Great Britain

American Staffordshire terrier

HEIGHT
17–19 in. / 43–48 cm

ORIGINAL FUNCTION
Bull baiting/fight dog

EXERCISE Much; long
walks, sport, tasks

HEALTH ISSUES Some

AFFECTION High

TRAINABILITY
Moderate/high

CARE NEEDS Low; teeth
cleaning if needed

WITH PETS Fair/good

WITH CHILDREN Fair/good

PROTECTIVITY
Moderate/high

FOR NOVICE OWNERS
Not suggested

Profile The ancestor of the American Staffordshire terrier was developed in England for bull and bear baiting and later for dogfights from a cross between old-style English bulldogs and various terriers. They are very close relatives of the American pit bull terrier that has recently been bred relatively independently from the "Amstaff." His tremendous strength and stocky body is unusual for his moderate height; he is extremely muscular, with a perfect balance of power and agility. His short, stiff, glossy coat can be of many color combinations.

Behavior and upkeep Modern "Amstaffs" are brave, docile, even-tempered, and make excellent guardians as well as great sport dogs. They are very affectionate with their family. However, they need extensive socialization to other dogs early in life. Their training requires firm, consistent, but not harsh discipline, as physical punishment may well lead to aggressive or anxious tendencies.

Nowadays, the popularity of the breed has led to indiscriminate breeding, and uneven temperament is not rare. Since they need plenty of exercise and guidance, they are best suited for owners who have the time, energy, and experience they need.

COUNTRY OF ORIGIN USA

Labrador retriever

HEIGHT
22–23 in./54–57 cm

ORIGINAL FUNCTION
water dog / retriever

EXERCISE much;
long walks, sport, tasks

HEALTH ISSUES some;
hip/elbow dysplasia,
cancer, melanomas,
soft-tissue sarcomas

AFFECTION high

TRAINABILITY high

CARE NEEDS low

WITH PETS excellent

WITH CHILDREN excellent

PROTECTIVITY low

FOR NOVICE OWNERS
excellent

COUNTRY OF ORIGIN Great Britain

Profile In recent years this (almost) perfect family dog has become the world's most popular breed. Their retrieving behavior was established when they worked on fishing boats in Newfoundland and had to grab the large hooked fish or pull the net ashore. This strongly built, well-balanced dog's distinctive features are the dense, water-resistant coat, the "otter" tail that is very thick toward the base, and the incredible swimming abilities. The coat can be black, yellow, or liver.

Behavior and upkeep Coupling companionship with the skills of a good sporting dog, this good-tempered and versatile breed makes a remarkable pet for active owners. Labradors are outgoing and very active, with a keen love of water. Their natural working instinct can show itself not just at field trials, but also during search-and-rescue tasks or as a guide dog. During the first year especially, they can be rather boisterous, drag the owner along on the leash, jump all over visitors, and even knock people over.

Bernese mountain dog

HEIGHT
24–28 in./60–70 cm

ORIGINAL FUNCTION
farm dog

EXERCISE moderate;
several short walks

HEALTH ISSUES many;
musculoskeletal problems
(dysplasia, arthritis), cancer,
gastric torsion, entropion,
progressive retinal atrophy

AFFECTION moderate/high

TRAINABILITY moderate

CARE NEEDS low/moderate;
regular brushing (heavy
shedding)

WITH PETS excellent

WITH CHILDREN excellent

PROTECTIVITY low

FOR NOVICE OWNERS
excellent

Profile This robust dog was developed to herd cattle, pull carts, and be a watchdog on the farmlands of Switzerland. No longer bred for working purposes, these attractive and good-mannered dogs are now kept as family pets. The Bernese stole the heart of dog fans with his bearlike appearance and flashy, tricolor coat with symmetrical white markings on the throat and chest. Unfortunately, partly because of health issues due to his popularity, this is one of the shortest-lived dog breeds.

Behavior and upkeep As with most large breeds, the Bernese is slow to mature; he may remain puppyish for some time. Bernese puppies should be socialized to accept new people and new situations to be self-confident adults. They are calm but social, typically placid with strangers, though some can be aloof, but never aggressive.

COUNTRY OF ORIGIN Switzerland

Hungarian vizsla

HEIGHT
22–24 in./54–60 cm

ORIGINAL FUNCTION
gundog

EXERCISE much;
long walks, sport, tasks

HEALTH ISSUES few:
allergies, epilepsy,
hip dysplasia

AFFECTION high

TRAINABILITY high

CARE NEEDS low;
trimming of nails if needed

WITH PETS good

WITH CHILDREN good

PROTECTIVITY low

FOR NOVICE OWNERS good

Profile This elegant and athletic pointer-type versatile gundog is famous for his excellent scenting powers. His russet gold color is very distinctive. Beside the popular short-haired version, there is a rather rare, more robust wirehaired variant (a recognized breed) that was created by crossbreeding with the German wirehaired pointer. In countries where tail docking is not prohibited by law, the tail may be shortened by one-quarter to avoid hunting hazards.

Behavior and upkeep Because of his adaptability and easygoing nature, the vizsla can be kept as an affectionate companion for any active family who can provide him with the exercise and attention he needs. Close contact with the owner is an absolute necessity for them. Their tendency to express their emotions, vocalizing diversely, is distinctive. Their intelligence and gentle, sensitive nature make them easy to train, although they can be overly shy.

COUNTRY OF ORIGIN Hungary

English springer spaniel

HEIGHT ~20 in./51 cm

ORIGINAL FUNCTION hunting dog

EXERCISE moderate; long walks, tasks

HEALTH ISSUES some; hip dysplasia, eye problems

AFFECTION high

TRAINABILITY high

CARE NEEDS moderate; weekly brushing, care of ears

WITH PETS good

WITH CHILDREN excellent

PROTECTIVITY low

FOR NOVICE OWNERS excellent

COUNTRY OF ORIGIN Great Britain

Profile Springers were named for their primary function—to frighten the game and make it "spring" out of hiding. Once, springer and cocker spaniels were often born in the same litter, being differentiated only by size. The English springer has a similar but smaller and less passionate cousin, the Welsh springer spaniel. In recent decades the English springer has split in two distinctive directions: The hunting type and the pet/show type are remarkably different. His dense, waterproof coat comes in three color patterns: black and white, liver and white, and tricolor. Nowadays, the popularity of the breed has resulted in some very uneven temperaments, aggression, and rage issues in addition to health concerns.

Behavior and upkeep The charming personality of springers, their smiling faces, and constantly wagging tails make them excellent companions for active families. They are full of energy, enjoy long walks and hikes, but prefer activities where they need to exercise their minds as well as their bodies. Separation anxiety is a common trait.

Poodle

HEIGHT toy: 9–11 in./23–28 cm; miniature: 11–14 in./28–35 cm; medium: 14–18 in./35–45 cm; standard: 18–24in./45–62 cm

ORIGINAL FUNCTION wildfowling

EXERCISE much; long walks, sport, tasks

HEALTH ISSUES some; epilepsy, hip dysplasia, allergies

AFFECTION high

TRAINABILITY high

CARE NEEDS moderate/high; weekly brushing, regular clipping, checking ears, trimming nails

WITH PETS excellent

WITH CHILDREN excellent

PROTECTIVITY low

FOR NOVICE OWNERS excellent (except toy)

COUNTRY OF ORIGIN France

Profile Originating in Germany and France, the poodle was developed to hunt and retrieve waterfowl. Because of his intelligence and friendliness, the poodle quickly became very popular as a companion dog. The four sizes (standard, medium, miniature, and toy) and range of colors still offer potential owners a wide assortment to choose from. Their characteristic frizzy, curly coat may come in colors of black, white, silver, brown, apricot, or cream. They can be trimmed or shaved for many different styles, some of which may seem exaggerated for most pet owners. Although they do not shed, the poodle's dense and curly coat requires immense care.

Behavior and upkeep Poodles have a joyful and loyal character; they are intelligent and elegant, outgoing and playful. The standard poodle is the calmest: stable, affectionate, and sometimes reserved. Toy poodles can be oversensitive or nervous without proper early training or if they come from unstable lines. All sizes enjoy and excel at dog sports such as agility and competitive obedience.

Australian shepherd

HEIGHT
18–23 in./46–58 cm

ORIGINAL FUNCTION
herding dog

EXERCISE much; long
walks, sport, tasks

HEALTH ISSUES some;
hip dysplasia, deafness,
epilepsy, eye problems

AFFECTION high

TRAINABILITY high

CARE NEEDS moderate;
regular brushing

WITH PETS excellent

WITH CHILDREN good

PROTECTIVITY
low/moderate

FOR NOVICE OWNERS good

Profile Originally bred to herd livestock, the Aussie, as it is nicknamed, is still an energetic working dog. Despite his name, the breed originated in the Unites States, where his history has been connected to Western horse shows and rodeos. Aussies have a muscular, well-balanced body, with picturesque coloring that offers variety and individuality. Their thick, water-resistant coat can be blue merle, black, red merle, or red—all with or without white and/or tan markings. The eyes must be fully surrounded by color (too much white is associated with a proneness to deafness), and can be blue, amber or include flecks and marbling. The tail can be naturally long or naturally short.

Behavior and upkeep Though bred to be a hardworking herding dog, Aussies do fine in cities if they get enough brain stimulation and exercise. They are loyal to their family but often standoffish with strangers. Since they can be sensitive and show fear-related aggression, they need a loving but confident owner. They are highly intelligent, eager to please their trainer, and want to be included in all activities.

COUNTRY OF ORIGIN USA

Golden retriever

HEIGHT
20–24 in./51–61 cm

ORIGINAL FUNCTION
gundog/retriever

EXERCISE moderate/much;
long walks, tasks

HEALTH ISSUES some; hip
dysplasia, allergies, heart
disease, epilepsy, cancer

AFFECTION high

TRAINABILITY
moderate/high

CARE NEEDS moderate;
weekly brushing

WITH PETS excellent

WITH CHILDREN excellent

PROTECTIVITY low

FOR NOVICE OWNERS
excellent

COUNTRY OF ORIGIN Great Britain

Profile Originating in the Scottish Highlands, the golden retriever was developed by crossing two now-extinct varieties of hunting dogs: the yellow flat-coated retriever and the Tweed water spaniel. Although originally bred as an all-round hunting dog, his happy look and sweet nature contributed to the breed's rapid rise in popularity as a pet. Golden retrievers should be strong, sturdy, and athletic rather than squab because of being fat. The color of the long, waterproof, double coat ranges from light cream to a rich gold. This breed is one of the most popular, and this has led to indiscriminate breeding practices resulting in puppies with unstable temperaments and long-term health issues.

Behavior and upkeep The golden retriever is popular not only as a pet but also for obedience competitions, hunting, and drug detection work. They are cheerful, well mannered with everyone after just minimal training, and forgiving of most mistakes made by inexperienced owners. Exuberant jumping and mouthiness, especially when young, can cause problems.

Belgian shepherd

HEIGHT
22–26 in./56–66 cm

ORIGINAL FUNCTION
herding dog

EXERCISE much; long
walks, sport, tasks

HEALTH ISSUES few;
epilepsy may occur
in some lines

AFFECTION high

TRAINABILITY high

CARE NEEDS low/moderate;
brushing when sheds

WITH PETS good

WITH CHILDREN good

PROTECTIVITY
moderate/strong

FOR NOVICE OWNERS
good (except Malinois)

Profile The Belgian shepherd is recognized in four distinct varieties: the long-haired, black Groenendael; the long-haired, fawn Tervueren; the short-haired, fawn Malinois; and the rough-haired, fawn, rather rare Laekenois. All variants are harmoniously proportioned, combining elegance and power. Even the long, abundant hair of the Tervueren and Groenendael does not need much grooming due to the almost "auto-shed" coat type. Although the four varieties are officially claimed to be distinct only in the color and length of their hair, only the Malinois is (and can be) used as a police and military dog.

Behavior and upkeep They can have a pleasurable life without herding sheep, if kept as a working dog or an active companion. They are very affectionate, highly intelligent, and trainable, and they do well in most dog sports. Being extremely devoted, they are not for owners who are away from home for long periods of time.

COUNTRY OF ORIGIN Belgium

Siberian husky

HEIGHT
20–24 in./50.5–60 cm

ORIGINAL FUNCTION
sled dog

EXERCISE moderate/much;
long walks, sport

HEALTH ISSUES few;
dysplasia, eye problems

AFFECTION moderate

TRAINABILITY low

CARE NEEDS low/moderate;
weekly brushing

WITH PETS fair

WITH CHILDREN good

PROTECTIVITY low

FOR NOVICE OWNERS
not suggested

COUNTRY OF ORIGIN USA

Profile This athletic, sled-pulling dog, capable of traveling great distances in the harsh Siberian tundra, is smaller and faster than the Alaskan malamute. The husky came to America across the Bering Strait more than a hundred years ago and quickly became one of the most popular breeds among Alaskan dog mushers. The thick double coat comes in many colors including various shades of gray and silver, sand, red, and black and white. The eyes can be brown or striking blue. Some huskies have a nose that has pink streaks, referred to as "snow nose." Similarly to wolves, they howl rather than bark. This hardy and healthy breed is claimed to be naturally clean and odor free.

Behavior and upkeep Siberian huskies are agile and free-spirited, and have a strong desire to roam. They are curious, independent-thinking, and not easy to train so it can be problematic for a first-time owner to manage them. In addition to long walks, they need to run and play with other dogs, so are best suited for families who have the time and energy to devote to their wellbeing.

Cocker spaniel

HEIGHT
15–16 in./38–41 cm

ORIGINAL FUNCTION
flushing dog

EXERCISE moderate;
long walks

HEALTH ISSUES some;
cardiovascular conditions,
skin disorders,
immunological disorders,
nephropathy, rage syndrome

AFFECTION moderate/high

TRAINABILITY moderate

CARE NEEDS moderate;
brushing the feathers
and checking ears
every other day

WITH PETS excellent

WITH CHILDREN
good/excellent

PROTECTIVITY low

FOR NOVICE OWNERS good

Profile The name "cocker" comes from their use to hunt woodcock. This English breed is distinguished from his New World cousin, the American cocker spaniel, which is smaller, has longer hair, and a distinctive abrupt stop on the head. The English cocker is a small to medium-sized dog with long ears, round heads, and a feathered coat. Cockers come in many colors—solid black, liver, red/gold, black and tan, liver and tan, particolor (with white), or different roans. Their extremely long, heavy ears are prone to ear infections.

Behavior and upkeep Cockers make great companion dogs for people who can give them the devotion and exercise they need. However, they can develop separation anxiety characterized by barking or destructive behavior. Considering the genetic background of some aggression issues in this breed, it is important to adopt a cocker, especially one of a solid color, from a reputable breeder. Cockers tend to beg for food with big, expressive eyes but gain weight easily, so owners should show great restraint and not overfeed them.

COUNTRY OF ORIGIN Great Britain

Shetland sheepdog

HEIGHT
13–16 in./33–40 cm

ORIGINAL FUNCTION
herding dog/companion

EXERCISE moderate;
long walks, tasks

HEALTH ISSUES some;
hip dysplasia, collie eye
anomaly, hemophilia,
congenital deafness, eye
diseases, seizures

AFFECTION high

TRAINABILITY high

CARE NEEDS moderate;
weekly brushing

WITH PETS good

WITH CHILDREN good

PROTECTIVITY low

FOR NOVICE OWNERS
good/excellent

Profile This small sheepdog was developed by farmers in the Shetland Islands, where his original function was to herd sheep. Their double, long, and dense coat can be black, sable, and blue merle, marked with varying amounts of white and/or tan. Being a very popular breed, there are many poorly bred puppies for sale, so it is important to find a reliable breeder who prioritizes both health and sound temperament.

Behavior and upkeep Shelties are smart and highly trainable, gentle and affectionate. Most of them are sensitive and do not respond well to harsh discipline. They are agile and make champions in obedience and agility, but they will be relatively inactive indoors and suit apartment living well. They tend to bark if left alone for long. They are affectionate with family and reserved or even shy with strangers, but never aggressive. Shelties are naturally willing, obedient, and kind; they can be excellent companions for elderly persons if they are walked daily.

COUNTRY OF ORIGIN Great Britain

Boxer

HEIGHT
21–25 in./53–63 cm

ORIGINAL FUNCTION bull
baiting/boar hunting

EXERCISE much/moderate;
short/long walks, in hot
weather more tasks than
physical exercise

HEALTH ISSUES many;
overheating, cancer,
hip dysplasia

AFFECTION high

TRAINABILITY
moderate/high

CARE NEEDS low/moderate;
wiping eyes and wrinkles
regularly

WITH PETS good

WITH CHILDREN excellent

PROTECTIVITY moderate

FOR NOVICE OWNERS good

Profile The boxer has been developed from the smaller Brabant bullenbeisser that was used to seize and hold the game during hunts. Today's boxer is a sturdy, strong, but also athletic and elegant dog with lively and powerful movement. The color is usually fawn or brindle with a black mask and with or without white markings. Boxers adapt well to apartment living but they are prone to drooling, flatulence, and snoring loudly. They are sensitive to both hot and cold weather.

Behavior and upkeep Contrary to what his facial expression suggests, the boxer primarily is a gentle and playful companion. He can get along well with other pets, including cats, if socialized to them at an early age. Undemanding, self-confident, calm, and brave, he is equally appreciated as a pet and as a guard. Boxers are easy to train due to their willingness to obey, their devotion and courage.

COUNTRY OF ORIGIN Germany

German Shepherd dog

HEIGHT
22–26 in./55–65 cm

ORIGINAL FUNCTION
herding dog

EXERCISE moderate/high;
long walks, tasks

HEALTH ISSUES some;
allergies, hip dysplasia,
loose hocks, epilepsy,
heart diseases, keratitis,
hemophilia

AFFECTION high

TRAINABILITY high

CARE NEEDS low/moderate;
weekly brushing

WITH PETS good

WITH CHILDREN
good/excellent

PROTECTIVITY
moderate/high

FOR NOVICE OWNERS good

Profile The German shepherd dog, also known as the Alsatian, was created from sheepdogs used in central and southern Germany at the end of the nineteenth century. He is still an agile, intelligent, and alert working dog with an optimal combination of stability, trainability, and the tendency to accept several handlers as a police dog. During recent decades, the tendency toward over-angulation of the hindquarters has resulted in a less square body shape and sloping back line. Colors are black with reddish-brown/yellow/light-gray markings; solid black; or grayish with darker shading (sable), and always with a black mask. The white color is not accepted.

Behavior and upkeep Modern German shepherd dogs can be very efficient guard dogs or excellent police dogs. From different lines, individuals can make reliable search-and-rescue dogs, guide dogs, and loving family companions. They are courageous and self-confident without being unnecessarily aggressive.

COUNTRY OF ORIGIN Germany

Great Dane

HEIGHT
28–35 in./72–90 cm

ORIGINAL FUNCTION
hunting

EXERCISE moderate;
long walks

HEALTH ISSUES many;
hip and elbow dysplasia,
varieties of heart diseases,
gastric torsion, bone cancer

AFFECTION high

TRAINABILITY moderate

CARE NEEDS low; tendency
to drool (wipe saliva)

WITH PETS excellent

WITH CHILDREN excellent

PROTECTIVITY
low/moderate

FOR NOVICE OWNERS good

COUNTRY OF ORIGIN Germany

Profile The breed has not much to do with Denmark, and they are called Great Danes only in English-speaking countries. Originally bred to hunt wild boar and even bears, this huge yet well-balanced and elegant dog comes in six colors—fawn, brindle, blue, black, harlequin, and mantle. A young Great Dane should not jump or run too long or fast, to minimize the stress on the growing bones and joints. Unfortunately, the Great Dane is not a long-lived breed, with an average life expectancy of 8–10 years or less. It is key to buy a Great Dane only from a reputable breeder who takes care about the health of the bloodline and makes sure that the parents are free of genetic diseases.

Behavior and upkeep In contrast to his size and mastiff-like appearance, the Great Dane is one of the most gentle and responsive breeds. They are tolerant playmates for children but due to their size they can accidentally knock them over. Some can be protective, but they can also be sensitive, so early socialization to strangers and unfamiliar dogs and habituation to weird sounds and situations is needed.

Rottweiler

HEIGHT
22–27 in./56–68 cm

ORIGINAL FUNCTION
herding/cart pulling

EXERCISE moderate/much;
long walks, tasks

HEALTH ISSUES many;
hip dysplasia, entropion,
ectropion, heart problems,
osteosarcoma, gastric
torsion, allergies

AFFECTION high

TRAINABILITY
moderate/high

CARE NEEDS low

WITH PETS fair/good

WITH CHILDREN good

PROTECTIVITY high

FOR NOVICE OWNERS
not suggested

COUNTRY OF ORIGIN Germany

Profile Originally bred in Germany to drive the herds of cattle to market and pull carts for butchers, rottweilers also defended their masters and their property. The strength and courage that allowed them to easily control and guard cattle later made them efficient police dogs. The rottweiler's outlook still embodies his qualities as a large, powerful guard dog. The black coat with clearly defined rich-tan markings lends him an even more threatening appearance. The tail should be kept in its natural condition, not docked.

Behavior and upkeep His watchfulness and self-assured courage as a defender are well known, but this also means that the rottie is not a dog for people who lack self-confidence. Despite their reputation as an aggressive breed, the rottie is a devoted family dog. Aggression toward other dogs is the biggest issue especially in the case of intact males. Kept as companions, they have a tendency to overeat and can gain weight. They do not drool, but many snore.

Border collie

HEIGHT 20-22 in./51-56 cm	**TRAINABILITY** moderate/high
ORIGINAL FUNCTION herding dog	**CARE NEEDS** low/moderate; weekly brushing
EXERCISE much; long walks, sport, tasks	**WITH PETS** excellent
HEALTH ISSUES few; congenital deafness, epilepsy, hip dysplasia, eye disorders	**WITH CHILDREN** good/excellent
	PROTECTIVITY low
AFFECTION high	**FOR NOVICE OWNERS** good

Profile The border collie is an athletic and acrobatic sheepdog, showing agility and maneuverability. The breed was developed on the highlands bordering Scotland, Wales, and England for working long hours on rugged terrain. They have been bred for performance, not appearance; their ears can be erect or semi-erect; they come in two coat varieties—moderately long and smooth—and they may come in just about any color and color pattern including solid, bicolor, tricolor, and merle.

Behavior and upkeep Border collies are highly intelligent and eager to please, and typically dominate the field in various canine sports, including obedience, flyball, and agility. However, their compulsion to herd can become misdirected unless they are properly exercised. Without a lot of physical and mental activity they may develop anxiety and behavior problems, such as separation anxiety, aggression or excessive vocalization.

COUNTRY OF ORIGIN Great Britain

Appendices

Bibliography

BOOKS

BEKOFF, M. (2007) *The emotional lives of animals: a leading scientist explores animal joy, sorrow, and empathy—and why they matter.* New Word Library, California.

BRADSHAW, J. (2011) *In Defence of Dogs.* Allen Lane, London.

CANDLAND, D.K. (1993) *Feral Children and Clever Animals.* Oxford University Press, New York.

COPPINGER, R., COPPINGER, L. (2002) *Dogs.* Chicago University Press, Chicago.

CSÁNYI, V. (2005) *If Dogs Could Talk.* North Point Press. New York.

DUGATKIN, L.E., TRUT, L. (2017) *How to Tame a Fox.* Chicago University Press, Chicago.

GRAMBO, R., COX, D (2015) *Wolf: Legend, Enemy, Icon.* Firefly Books.

HARE, B., WOODS, W. (2013) *The Genius of Dogs.* Peguin Book, New York.

HOROWITZ, A. (2009) *Inside of a Dog.* Scribner, New York.

HOROWITZ, A., FRICKE, W. (2007) *Anatomy of the Dog: An Illustrated Text,* Schluetersche Publisher, Hannover.

KAMINSKI, J., MARSHALL-PESCINI, S. (2014) *The Social Dog.* Academic Press, San Diego.

MECH, D.L. (1870/2012) *Wolf: The ecology and behavior of an Endangered Species.* Natural History Press, New York.

MECH, D.L., BOITANI, L. (2007) *Wolves: Behavior, Ecology, and Conservation.* University Chicago Press, Chicago.

MIKLÓSI, Á. (2014) *Dog Behaviour, Evolution and Cognition.* Oxford University Press, Oxford.

MOREY, D.F. (2010) *Dogs. Domestication and the Development of a Social Bond.* Cambridge University Press, Cambridge.

MUSIANI, M., BOITANI, L. (2010) *The World of Wolves: New Perspectives on Ecology, Behavior, and Management (Energy, Ecology and Environment)* University of Calgary Press, Calgary.

OSTRANDER, E.A., RUVINSKY, A. (2012). *The Genomics of the Dog.* CABI Publishing, Wallingford.

PETERSON, B. (2017) *Wolf Nation: The Life, Death,* and *Return of Wild American Wolves.* Da Capo Press, Boston.

PILLEY, J.W., HINZMANN, H. (2014) *Chaser: Unlocking the Genius of the Dog Who Knows a Thousand Words.* Oneworld Publications, London.

REECE, W.O., ROWE, E.W. (2017) *Functional Anatomy and Physiology of Domestic Animals.* John Wiley, Hoboken.

SERPELL, J. (2017). *The Domestic Dog.* Cambridge University Press, Cambridge.

SCOTT, J.P., FULLER, J.L. (1974/1997) *Genetics and the Social Behavior of the Dog.* Chicago University Press, Chicago

SHELDON, J.W., (1988) *Wild Dogs: The Natural History of the Nondomestic Canidae.* Academic Press, San Diego.

STILWELL (2016) *The Secret Language of Dogs: Unlocking the Canine Mind for a Happier Pet.* Crown Publishing, New York

WAND, X., TEDFORD, R.H. (2008) *Dogs. Their fossil relatives and evolutionary history.* Columbia University Press, New York

YIN, S. (2010) *How to Behave So Your Dog Behaves.* TFH Publications, Inc.

JOURNALS

ANDICS, A., GÁBOR, A., GÁCSI, M., FARAGÓ, T., SZABÓ, D., MIKLÓSI, Á. (2016) Neural mechanisms for lexical processing in dogs. *Science*, 353: 1030-1032.

BROWN, S.W., GOLDSTEIN, L.H. (2011) Can Seizure-Alert Dogs predict seizures? *Epilepsy Research* 97, 236–42.

CUSTANCE, D., MAYER, J. (2012) Empathic-like responding by domestic dogs (*Canis familiaris*) to distress in humans: an exploratory study. *Animal Cognition* 15, 851–9.

DUFFY, D.L., HSU, Y., SERPELL, J. (2008) Breed differences in canine aggression. *Applied Animal Behaviour Science* 114, 441–60.

FISET, S., LEBLANC, V. (2007) Invisible displacement understanding in domestic dogs (*Canis familiaris*): the role of visual cues in search behavior. *Animal Cognition* 10, 211–24.

FUGAZZA, C., MIKLÓSI, Á. (2015) Social learning in dog training: the effectiveness of the Do as I do method compared to shaping/clicker training. *Applied Animal Behaviour Science*, 171: 146-151.

FUGAZZA, C., POGÁNY, Á., MIKLÓSI, Á. (2016) Recall of Others' actions after incidental encoding reveals episodic-like memory in dogs. *Current Biology*, 26, 3209-3213.

GÁCSI, M., MCGREEVY, P., KARA, E., MIKLÓSI, Á. (2009) Effects of selection for cooperation and attention in dogs. *Behavioral and Brain Functions*, 5: 31.

GÁCSI, M., MAROS, K., SERNKVIST, S., FARAGÓ, T., MIKLÓSI, Á. (2013) Human analog safe haven effect of the owner: behavioral and heart rate response to stressful social stimuli in dogs. *PLoS ONE*, 8: e58475.

GAUNET, F. (2010) How do guide dogs and pet dogs (*Canis familiaris*) ask their owners for their toy and for playing? *Animal Cognition* 13, 311–23.

GOSLING, S.D., KWAN, V.S.Y., JOHN, O.P. (2003) A dog's got personality: a cross-species comparative approach to personality judgments in dogs and humans. *Journal of Personality and Social Psychology* 85, 1161–9.

HALL, N.J., WYNNE, C.D.L. (2012) The canid genome: behavioral geneticists' best friend? *Genes, Brain, and Behavior* 11, 889–902.

HARE, B., TOMASELLO, M. (2005a) Human-like social skills in dogs? *Trends in Cognitive Sciences* 9, 439–44.

HOROWITZ, A. (2009) Disambiguating the 'guilty look': salient prompts to a familiar dog behaviour. *Behavioural Processes* 81, 447–52.

HUBER, L., RACCA, A., SCAF, B. et al. (2013) Discrimination of familiar human faces in dogs (*Canis familiaris*). *Learning and Motivation* 44, 258–69.

KAMINSKI, J., NEUMANN, M., BRÄUER, J. et al. (2011) Dogs, *Canis familiaris*, communicate with humans to request but not to inform. *Animal Behaviour* 82, 651–8.

KAMINSKI, J., PITSCH, A., TOMASELLO, M. (2013) Dogs steal in the dark. *Animal Cognition* 16, 385–94.

KUBINYI, E., PONGRÁCZ, P., MIKLÓSI, Á. (2009) Dog as a model for studying con- and hetero-specific social learning. *Journal of Veterinary Behavior*, 4: 31-41.

KUBINYI, E., VAS, J., HÉJJAS, K., RONAI, ZS., BRÚDER, I., TURCSÁN, B., SASVÁRI-SZÉKELY, M., MIKLÓSI, Á. (2012) Polymorphism in the tyrosine hydroxylase (TH) gene is associated with activity-impulsivity in German Shepherd dogs. *PLoS One*, 7: e30271.

KUKEKOVA, A.V., TEMNYKH, S.V., JOHNSON, J.L. et al. (2012) Genetics of behavior in the silver fox. *Mammalian Genome* 23, 164–77.

LI, Y., VONHOLDT, B.M., REYNOLDS, A., BOYKO, A.R., WAYNE, R.K., WU, D.D., ZHANG, Y.P. (2013) Artificial selection on brain-expressed genes during the domestication of the dog. *Molecular Biology and Evolution* 8, 1867–76.

MAROS, K., PONGRÁCZ, P., BÁRDOS, GY., MOLNÁR, CS., FARAGÓ, T., MIKLÓSI, Á. (2008) Dogs can discriminate barks from different situations. *Applied Animal Behaviour Science*, 114: 159–167.

MCGREEVY, P.D., Nicholas, F.W. (1999) Some practical solutions to welfare problems in dog breeding. *Animal Welfare* 8, 329–41.

MCGREEVY, P.D., STARLING, M., BRANSON, N.J. et al. (2012) An overview of the dog-human dyad and ethograms within it. *Journal of Veterinary Behavior: Clinical Applications and Research* 7, 103–17.

MECH, L.D. (1999) Alpha status, dominance, and division of labor in wolf packs. *Canadian Journal of Zoology* 77, 1196–203.

MEROLA, I., PRATO-PREVIDE, E., MARSHALL-PESCINI, S. (2012) Social referencing in dog-owner dyads? *Animal Cognition* 15, 175–85.

MIKLÓSI, Á., TOPÁL, J. (2013) What does it take to become "best friends"? Evolutionary changes in canine social competence. *Trends in Cognitive Sciences*, 17: 287-294.

MIKLÓSI, A., KUBINYI, E. (2016) Current trends in Canine problem-solving and cognition. *Current Directions in Psychological Science*, 25: 300–306.

MILLS, D.S. (2005) What's in a word? A review of the attributes of a command affecting the performance of pet dogs. *Anthrozoös* 18, 208–21.

MONGILLO, P., ARAUJO, J.A., PITTERI, E. et al. (2013a) Spatial reversal learning is impaired by age in pet dogs. *Age (Dordrecht, Netherlands)* 35, 2273–82.

NAGASAWA, M., KIKUSUI, T., ONAKA, T., OHTA, M. (2009) Dog's gaze at its owner increases owner's urinary oxytocin during social interaction. *Hormones and Behavior* 55, 434–41.

POLGÁR, Z., MIKLÓSI, Á., GÁCSI, M. (2015) Strategies used by pet dogs for solving olfaction-based problems at various distances. *PLoS ONE*, 10: e0131610.

PONGRÁCZ, P., MOLNÁR, Cs., DÓKA, A., MIKLÓSI, Á. (2011) Do children understand man's best friend? Classification of dog barks by pre-adolescents and adults. *Applied Animal Behaviour Science*, 135: 95-102.

RAMOS, D., ADES, C. (2012) Two-item sentence comprehension by a dog (*Canis familiaris*). *PLoS ONE* 7, e29689.

RANGE, F., HORN, L., VIRANYI, Z., HUBER, L. (2009a) Theabsence of reward induces inequity aversion in dogs. *Proceedings of the National Academy of Sciences of the United States of America* 106, 340–5.

RANGE, F., VIRANYI, Zs., HUBER, L. (2007) Selective imitation in domestic dogs. *Current Biology*, 17: 868-872.

SAVOLAINEN, P., ZHANG, Y., LUO, J. et al. (2002) Genetic evidence for an East Asian origin of domestic dogs. *Science*, 298, 1610–13.

SVARTBERG, K. (2006) Breed-typical behaviour in dogs—Historical remnants or recent constructs? *Applied Animal Behaviour Science* 96, 293–313.

SZABÓ, D., GEE, N. R., MIKLÓSI, Á. (2016) Natural or pathologic? Discrepancies in the study of behavioral and cognitive signs in aging family dogs. *Journal of Veterinary Behavior: Clinical Applications and Research*, 11, 86-98.

TOPÁL, J., GERGELY, Gy., ERDŐHEGYI, Á., CSIBRA, G., MIKLÓSI, Á. (2009) Differential sensitivity to human communication in dogs, wolves, and human infants. *Science*, 325: 1269-1272.

TURCSÁN, B., RANGE, F., VIRÁNYI, Zs., MIKLÓSI, Á., KUBINYI, E. (2012) Birds of a feather flock together? Perceived personality matching in owner–dog dyads. *Applied Animal Behaviour Science*, 140: 154-160.

UDELL, M.A.R., DOREY, N.R., WYNNE, C.D.L. (2011) Can your dog read your mind? Understanding the causes of canine perspective taking. *Learning & Behavior* 39, 289–302.

VONHOLDT, B.M., POLLINGER, J.P., LOHMUELLER, K.E. et al. (2010) Genome-wide SNP and haplotype analyses reveal a rich history underlying dog domestication. *Nature* 464, 898–902.

WARD, C., BAUER, E.B., SMUTS, B.B. (2008) Partner preferences and asymmetries in social play among domestic dog, Canis lupus familiaris, littermates. *Animal Behaviour* 76, 1187–99

WAYNE, R.K., VON HOLDT, B.M. (2012). Evolutionary genomics of dog domestication. *Mammalian Genome* 23, 3–18.

MAGAZINES

The Bark
www.thebark.com

Dogs Monthly
www.dogsmonthly.co.uk

The Whole Dog Journal
www.whole-dog-journal.com

WEBSITES

Family Dog Project
(Eötvös Loránd University, Hungary)
familydogproject.elte.hu

Clever Dog Lab
(University of Vienna, Austria)
cleverdoglab.at

Wolf Science Center
(Ernstbrunn, Austria)
wolfscience.at

Author Biographies

Ádám Miklósi is a professor and head of the Department of Ethology at Eötvös Loránd University in Budapest, Hungary. He is also the co-founder and leader of the *Family Dog Project*, a center for the study of human-dog interaction from an ethological perspective. Adam is the author of *Dog Behaviour, Evolution, and Cognition* (2nd edition 2014) published by Oxford University Press.

Tamás Faragó is a research fellow at the Comparative Ethology Research Group of the Hungarian Academy of Science at the Department of Ethology, Eötvös Loránd University. He studied vocal communication of dogs in his PhD work, with a special focus on the communicative aspects of dog growls. He has contributed to 25 papers and book chapters.

Claudia Fugazza is a post-doctoral research fellow at the Department of Ethology at Eötvös Loránd University, where she specializes in social learning and imitation in dogs. She has previously worked as a dog trainer and is the author of *Do as I Do; Using Social Learning to Train Dogs* (First Stone, 2008).

Márta Gácsi is senior researcher at the Comparative Ethology Research Group of the Hungarian Academy of Sciences and investigates dog-human interactions in the Family Dog Project. At Eötvös University she teaches cognitive ethology, evolution of communication, human-animal interaction, and evolution of canids. Marta is the author/co-author of more than 70 papers and book chapters on dog behavior.

Enikő Kubinyi is a senior research fellow at the Department of Ethology at Eötvös Loránd University, where her research focuses on the comparative analysis of dogs' and wolves' cognition, ethorobotics, personality, behavioral genetics, and cognitive aging in dogs. She has written more than 30 papers about the behavior of canines and other animals and also covers the topic in two blogs.

Péter Pongrácz is an associate professor in the Department of Ethology at Eötvös Loránd University, where he investigates dog-human interactions. His research focuses mainly on the acoustic communication of dogs and social learning. He teaches various ethology-related courses at the university and, with his students, has published nearly 80 peer-reviewed papers and book chapters.

József Topál was a founding member of the Family Dog Project and is currently vice director of the Institute of Cognitive Neuroscience and Psychology and the head of the Psychobiology Research Group, RCNS, HAS, Budapest. He has published extensively on dog behavior and dog-human interaction and is the author of more than 100 scientific publications.

Index ~⟋

Acknowledgments ⟋

The publisher would like to thank the following for permission to reproduce copyright material:

Aeroflot: 30L, 30R.

Alamy/AF Fotografie: 142TR; Blickwinkel: 73; Everett Collection: 136; franzfoto.com: 12; Heritage Image Partnership Ltd: 134B; Juniors Bildarchiv GmbH: 21, 84, 205, 207; Mark Scheuern: 174B; Susann Parker: 75R; Jose Luis Stephens: 146T; Jack Sullivan: 59; Tierfotoagentur: 33T, 105L; Joe Vogan: 19 (9).

Nora Bunford: 117B.

FLPA/ImageBroker: 29R.

Claudia Fugazza: 152TR.

Márta Gácsi: 29L, 31BL, 31BR, 47T, 56L, 99, 105C, 129TC, 139L, 139R.

Getty Images/Timothy A. Clary/AFP: 169T; Corbis: 142; De Agostini: 134T; Dohlongma - HL Mak: 91B; Ronan Donovan/National Geographic: 26T; Patrick Endres/Design Pics: 8; Hulton Archive: 137R; Ixefra: 148; Michelle Kelley: 77; Isaac Lawrence/AFP: 58T; David Leswick : 91T; Joe McDonald/ Corbis Documentary: 14T; Mediaphotos: 124; Tracey Morgan/Dorling Kindersley: 212; Laurence Mouton/Canopy: 149B; Kevin Oke/All Canada Photos: 142TL; Photonica World: 179TR; Universal History Archive: 214; Vanessa Van Ryzin, Mindful Motion Photography: 161; Visuals Unlimited: 18 (3); Westend61: 144.

Marc Henrie: 183, 186, 187, 188, 191, 192, 195, 196, 198, 199, 203, 204, 206, 208, 210, 213.

iStock/Eyecrave: 140TR; Huseyintuncer: 35; Nicoolay: 6, 143B; Roir88: 18 (1).

The Metropolitan Museum of Art: 24TR, 24C, 24BL, 24BR, 135.

Nature Picture Library/Eric Baccega: 114; Bartussek/ARCO: 37; Florian Mallers: 72; The Natural History Museum: 14B; Petra Wegner: 120, 121, 182, 184, 201; Solvin Zankl: 146B.

Marie-Lan Nguyen: 32.

Shutterstock/Adya: 176; Aekarin Kitayasittanart: 54TL; Africa Studio: 132, 165; Alberto Chiarle: 19 (7); alexei_tm: 78, 108L; Alexey Fursov: 160TR; Alexey Kozhemyakin: 175; Alexkatkov: 86B; Alis Leonte 10; Alzbeta: 129TL; Ammit Jack: 168; ANADMAN BVBA: 81B; Andrey Oleynik: 4B; Aneta Jungerova: 129TR; anetapics 89; Anna Hoychuk: 178C; Anton Gumen: 33B; ARENA Creative: 100T; ATSILENSE: 158BR; AustralianCamera: 26B; bbernard 127; Bildagentur Zoonar GmbH 81T; Boryana Manzurova 130; Bruno Rodrigues B Silva: 54TR; Budimir Jevtic: 140B; Charlene Bayerle: 153T; chittakom59: 82B; Choniawut: 54TC; Chris Fourle: 18 (2); Christian Mueller: 90; Cryber: 76; Csanad Kiss: 3TR, 49TL, 54CR, 115T, 200; cynoclub: 46L, 47L, 66L, 185; Dan Kosmayer: 153B; Dan Tautan: 25; Danny Iacob: 104; David Tadevosian: 157R; Daxiao Productions: 105R; Daz Stock: 178T; deepspace: 145; Degtyaryov Andrey: 179TL; dezi: 118; DGLimages: 141; Didkovska Ilona: 88L; Dmitry Pichugin: 137L; dominibrown: 116; Dora Zett: 47R, 62B, 95, 194, 202; Dorottya Mathe: 46CR; Dragos Lucian Birtoiu: 27R; Dusan Petkovic: 169B; eClick: 79B; egyjanek: 54CL; Ekaterina Brusnika: 119; elbud: 92; Eric Isselee: 2TL, 27L, 42, 46R, 80CL, 96C, 98, 101T, 129B, 131T, 131B, 158BL, 163T, 174TL, 179B, 181B, 190, 193, 211; Erik Lam 140CR; everydoghasastory: 123B; Evgeny Tomeev: 51TC; Ewais: 62T; FCSCAFEINE: 112; FiledIMAGE: 34B; Geir Olav Lyngfjell: 47CL; Gelpi: 43, 100B, 115B, 159, 209; George Lee: 108R; Golbay: 51TL; Goldution: 131L; Gonzalo Jara: 79CL; goodluz: 164; Grey Tree Studios: 158T; Grigorita Ko: 48, 85, 180T; Grisha Bruev: 44; Halfpoint: 157L; Hein Nouwens: 17L; Hurricanehank: 68; hyperborean-husky: 86T; Ivonne Wierink: 101B; Jagodka: front cover, 2, 64, 174TR; Jakkrit Orrasri: 160TC; James Kirkikis: 162; jlsphotos: 83TL, 83CL; Josef Pittner: 19 (9); Julia Shepeleva: 177; Kaja N: 129CL;

kirian: 49TR; Klaus Hertz-Ladiges: 170R; Ksenia Raykova: 126; Kukiat B: 111; Kuznetsov Alexey: 67B; larstuchel: 63R; leonardo2011: 93B; Life In Pixels: 163B; Lmfoto: 122; Lori Labrecque: 87; Luke23: 19 (6); Maria Sbytova: 151; markos86: 47CR; Melica: 60; Melle V: 149T; MF Photo: 56R; Michael Dorogovich: 74; michaelheim: 99T, 152TL; michaeljung: 152B; Mikkel Bigandt: 128; Milica Nistoran: 109; MirasWonderland: 181TL, 189; Morphart Creation: 4T, 49C; Nagel Photography: 16; Natalia Fedosova: 102; NaturesMomentsuk: 18 (5); Nikol Mansfeld: 79CR; Nikolai Tsvetkov 62 (BL); Nilkanth Sonawane: 34T; Olga; Ovcharenko: 45; Oquzum: 23; Osadchaya: Olga 7, 154; otsphoto: 38, 88R, 93T; outdoorsman: 65; PardoY: 40, 54B; Petr Schmid: 155; Photick: 110; Photology1971: 180B; Picsoftheday: 147; PixieMe: 160BL; Popova Valeriya: 71; r.classen: 3B, 58B; Ratikova: 160TL; Rebecca Ashworth: 5; Ricantimages: 66BR; Rosa Jay: 46L; RTimages: 181TR; santisuk; wonganu: 36T; Sergey Fatin: 51B; Sergiy Kuzmin: 51TR; Sigma S: 83TC, 83TR, 83CR; Simon Eeman: 70; Sirko Hartmann: 170L; SOPRADIT: 61; SpeedKingz: 75L, 150, 160BR; Stuart G Porter: 18 (4); StudioByTheSea: 36B; studiolaska: 140TL, 140TC, 140CL; Susan Schmitz: 172, 197; TatyanaPanova: 67T; thka: 129CR; Thomas Soellner: 163C; Timodaddy: 79TL; Utekhina Anna: 107B; Vera Zinkova: 79TR, 107T; Viktor Kholosha: 79TC; Vladimir Wrangel: 14C; whitehorseexotics: 178B; WilleeCole Photography: 66TR; Zayats Svetlana: 63L; Zuzule: 82T, 117T.

Science Photo Library/Patrick Llewellyn-Davies: 171; Mona Lisa Production: 166; Louise Murray: 123T; Sciepro: 53.

Roman Uchytel, prehistoric-fauna.com: 17.

Walters Art Museum: 33T (inset).

All reasonable efforts have been made to trace copyright holders and to obtain their permission for the use of copyright material. The publisher apologizes for any errors or omissions and will gratefully incorporate any corrections in future reprints if notified.